Decisions about Re-engineering

Decisions about Re-engineering

Briefings on Issues and Options

Bart O'Brien

CHAPMAN & HALL

London · Glasgow · Weinheim · New York · Tokyo · Melbourne · Madras

Published by Chapman & Hall,
2-6 Boundary Row, London SE1 8HN, UK

Chapman & Hall, 2-6 Boundary Row, London SE1 8HN, UK

Blackie Academic & Professional, Wester Cleddens Road,
Bishopbriggs, Glasgow G64 2NZ, UK

Chapman & Hall GmbH, Pappelallee 3, 69469 Weinheim, Germany

Chapman & Hall USA, 115 Fifth Avenue, New York, NY 10003, USA

Chapman & Hall Japan, ITP-Japan, Kyowa Building, 3F, 2-2-1 Hirakawacho,
Chiyoda-ku, Tokyo 102, Japan

Chapman & Hall Australia, Thomas Nelson Australia, 102 Dodds Street,
South Melbourne, Victoria 3205, Australia

Chapman & Hall India, R. Seshadri, 32 Second Main Road, CIT East, Madras
600 035, India

First edition 1995
© 1995 Bart O'Brien

ISBN 0 412 72300 X
Printed in Great Britain by St Edmundsbury Press, Bury St Edmunds, Suffolk

Henk and Walter Baghuis of Compubest BV, Rotterdam, the
Netherlands helped the author by generously providing PostScript
printing facilities.

A catalogue record for this book is available from the British Library
Library of Congress Catalog Card Number: 95-70020

∞ Printed on permanent acid-free text paper, manufactured in accordance with
ANSI/NISO Z39.48-1992 and ANSI/NISO Z39.48-1984 (Permanence of Paper)

Contents

Introduction

Re-engineering has been *the* management craze of the early nineties. Two influential articles appeared in 1990, and before long every management, IT and trade journal had published similar articles. At least 20 books on re-engineering were published in English in 1992-94, one of them a best-seller. Meanwhile, management consultancies began to promote re-engineering services. Business schools set up courses and even professorial chairs in re-engineering. By the end of 1994 a search on the keyword *re-engineering* in a bibliographic database would yield more than 2000 references.

When *re-engineering* eventually falls from favour, as all boardroom buzzwords do, how will anything be different? Like others before, it has generated hype and thus confusion and thus some bad management; nevertheless, at least *some* proponents of re-engineering have stimulated new insights into certain enduring issues in the design and introduction of systems in organisations. The challenge is to cut through the tangles of exhortation, platitude and glib assumption to extract the interesting issues and debate them.

Other books assert the importance of re-engineering and advise how to carry it out. This book too aims to promote successful re-engineering, but it operates on an extra plane: it contrasts and discusses the published ideas of other people. The ideas in (say) the breathlessly enthusiastic best-seller by Hammer and Champy are worth knowing about for at least two reasons. First, you may lose credibility with colleagues and business associates if you are not familiar with the mainstream themes and case-studies. More important, critical examination of the content of such a book is a good start towards thinking through some enduring issues in the design and introduction of systems, whose trickiness is likely to survive the vagaries of fashion.

It might seem that one essential precondition for any smash-hit management craze should be to have one definite name, but in practice the following terms are used more or less indiscriminately: 're-engineering', 'business re-engineering', 'business process re-engineering', 'process redesign', 'business process redesign', 'business process innovation' and 'process innovation'. This book sticks to 're-engineering' as far as possible.

▼ The demystification of re-engineering yields certain general principles of clear thinking, that can be equally serviceable in hacking through the thickets of words on any number of other management topics. Notes on this more general plane are made from time to time in text marked as this is. ▲

Articles and Books on Re-engineering

It is unusual to begin with a bibliography, but this book will devote considerable attention to the ideas about re-engineering that others have published.

Two influential journal articles about re-engineering appeared in 1990. Once the rage caught on, their authors followed up with influential books. Things moved quickly. By 1994 one of these authors was concerned enough about the uncritical adoption of re-engineering themes to write an article about some myths that had grown up.

Two Influential Articles

Davenport, TH and Short, JE. 'The New Industrial Engineering: Information Technology and Business Process Redesign', *Sloan Management Review*, Summer 1990, pp. 11-27.

Hammer, M. 'Reengineering Work; Don't Automate, Obliterate', *Harvard Business Review*, July-August 1990, pp. 104-112.

Other Interesting Articles

Bendall-Harris, V. 'Customer-focused Re-engineering in Telstra: Corporate Complaints Handling in Australia', *Business Change & Re-engineering*, Summer 1994, pp. 7-14.

Caron, JR et al. 'Business Reengineering at CIGNA Corporation: Experiences and Lessons Learned From the First Five Years', *MIS Quarterly*, September 1994, pp. 233-250.

Cavanaugh, HA. 'Re-engineering: Buzz word, or powerful new business tool?', *Electrical World*, April 1994, pp. 7-15.

Clemmer, J. 'Process re-engineering and process improvement: Not an either/or choice', *CMA Magazine*, June 1994, pp. 36-39.

Davenport, TH and Stoddard, DB. 'Reengineering: Business Change of Mythic Proportions?', *MIS Quarterly*, June 1994, pp. 121-127.

Davenport, TH, interviewed by Watts, J. *Business Change & Re-engineering*, Summer 1994, pp. 2-6.

Earl, MJ and Khan, B. 'How New is Business Process Redesign?', *European Management Journal*, March 1994, pp. 20-30. (Much the same article appeared in *Journal of Strategic Information Systems*, 1994 3(1), pp. 5-22.)

Edwards, C and Peppard, J. 'Forging a Link Between Business Strategy and Business Reengineering', *European Management Journal*, December 1994, pp. 407-18.

Hall, G et al. 'How to make Reengineering really work', *Harvard Business Review*, November-December 1993, pp. 119-131.

Hammer, M. 'Hammer defends re-engineering', *The Economist*, 5 November 1994, p. 80.

Keenan, W. 'If I Had A Hammer', *Sales & Marketing Management*, December 1993, pp. 56-61.

Lynch, D et al. 'Transforming the Performance of Melbourne Water', *Business Change & Re-engineering*, December 1994, pp. 19-32.

Proc, M et al.. 'The Premier of Ontario's Correspondence Unit: Strategic Redesign of the Communications Process', *Business Change & Re-engineering*, Winter 1993, pp. 7-14.

Short, JE and Venkatraman, N. 'Beyond Business Process Redesign: Redefining Baxter's Business Network', *Sloan Management Review*, Fall 1992, pp. 7-21.

Strassmann, PA. 'The Hocus-Pocus of Reengineering', *Across the Board*, June 1994, pp. 35-38.

Talwar, R. 'Business Re-engineering — a Strategy-driven Approach', *Long Range Planning*, December 1993, pp. 22-40.

'Return of the stopwatch', *The Economist*, 23 January 1993, p. 73.

'Take a clean sheet of paper', *The Economist*, 1 May 1993, pp. 71-2.

'The wonders of workflow', *The Economist*, 11 December 1993, p. 80.

Several thousand articles have been published in English on re-engineering since 1990. The selection above is strongly biased towards the more thoughtful — or, in a few cases, not all that thoughtful but still thought-provoking. The articles illustrate a range of attitudes: Caron et al and Hall et al don't question the main premises of re-engineering contained in the two influential articles but they develop some nuances; on the other hand, Davenport's two pieces come close to recantation of some points in his original article; and Strassmann likens proponents of re-engineering to 'political hijackers' such as Robespierre, Lenin and Mao.

Two Influential Books

Davenport, TH. *Process Innovation, Reengineering Work through Information Technology* (Harvard Business School Press, 1993)

Hammer, M and Champy, J. *Reengineering the Corporation, a Manifesto for Business Revolution* (Harper Business, 1993)

In the briefings that follow, these two books are often the starting-point for debate. Hammer and Champy's book has probably been read by more people than all the other books put together. Moreover, by the standards of the management best-seller with messianic title, it is a fairly cogent piece of work.

Davenport's book is less incisive than his articles. The ideas are sometimes muffled by Harvardspeak. Nevertheless, they are more thoughtful and realistic ideas than those in Hammer and Champy's book or any of the others listed here.

▼ In any subject, it is natural to want to know about good ideas, principles and techniques, rather than stupid or feeble ones. With (say) civil engineering that is the end of the story, but not with a subject like management. Here it is also worth knowing and thinking about *influential* ideas, whatever their quality. If you don't, you risk losing credibility with your peers. Moreover, if you are well-prepared beforehand, it is much easier to expose the shortcomings of any influential, seductive half-truths that may surface in debates with colleagues or clients. And again, thinking through exactly why a certain piece of hype seems too shallow to be taken seriously is an excellent start towards gaining genuine insight into the complexities of an issue. ▲

Other Books

Andrews, DC and Stalick, SK. *Business Reengineering, the Survival Guide* (Yourdon Press, 1994)

Carr, DK et al. *Breakpoint, Business Process Redesign* (Coopers & Lybrand, 1992)

Cross, KF et al. *Corporate Renaissance, The Art of Reengineering* (Blackwell Business, 1994)

Dur, RCJ. *Business Reengineering in Information Intensive Organizations* (from the author, PO Box 356, 2600 AJ Delft, the Netherlands, 1992)

Johansson, HJ et al. *Business Process Reengineering, Breakpoint Strategies for Market Dominance* (Wiley, 1993)

Morris, D, and Brandon, J. *Re-engineering Your Business* (McGraw-Hill, 1993)

Obeng, E, and Crainer, S. Making Re-engineering Happen (Financial Times and Pitman, 1994)

Petrozzo, DP and Stepper, JC. *Successful Reengineering* (Van Nostrand Reinhold, 1994)

Spurr, K et al (eds.). *Software Assistance for Business Re-engineering* (Wiley, 1993)

Towers, S. *Business Process Re-engineering, A Practical Handbook for Executives* (Stanley Thornes, 1994)

Business Process Reengineering, Current Issues and Applications (Institute of Industrial Engineers, Norcross, GA, USA, 1993)

These eleven books fall into three groups. *First*, four books present an enthusiastic overview of re-engineering themes. Carr et al and Johansson et al are all partners of Coopers & Lybrand, a consultancy company. The Carr et al book is little more than consultants' glossy-brochure material. Parts of the Johansson et al book read like a parody of the self-help, can-do management text, but some moderately detailed examples can be used as a starting-point for critical thinking.

The first one-third of the book by Obeng and Crainer is an excruciatingly naff dialogue between two re-engineering bores stuck with each other on a flight; the rest of the book is also very shallow. The brief, absurdly expensive book by Towers, offered as 'ideal for busy managers', is certainly an undemanding selection of clichés, but that does not make it worth reading.

It is difficult to see the point of books such as these four. They are no more profound than the Hammer and Champy work, which is better written and much more influential.

In the *second* group are four books that each describe a particular, standard, step-by-step approach to re-engineering. They provide more exhortation and platitude than penetrating analysis of the tricky issues. The first three books are treated fairly sternly in the briefings that follow, while that by Petrozzo and Stepper is mainly too superficial to be worth criticising. Of the four, the book by Cross et al is definitely the least bad.

There is a *third*, miscellaneous, group. Dur's book describes a method of modelling and simulating business processes. Its detailed example cases are interesting raw material for discussion of *any* modelling methods, not just Dur's. Spurr et al provide a guide to software products that facilitate the work of re-engineering. There is description but not comparison or classification or any other analytical technique for making sense of the market's variety. The IIE book is an anthology of more than 30 articles, most of them no better or worse than a thousand others. Still, the two influential articles are included.

▼ Practically any book on re-engineering can be worth reading if it contains plenty of detailed case-study material. Study that material to assess the merits or shortcomings in the book's ideas, and to test out other ideas of your own. This principle of the value of detailed example material applies to books on most aspects of management science. ▲

1.   Kinds of Decision

ISSUES

The rhetoric of re-engineering can be dangerous if accepted uncritically, but valuable as a stimulus to debate about certain enduring issues in the design and development of systems in organisations. A handy technique for moving beyond the fine words into thoughtful inquiry is to concentrate on the *decisions* that people take about re-engineering.

Defined in preliminary and not very precise terms, a re-engineering project seeks to make radical improvements in a certain chunk of an organisation's operations. Suppose, as a thought-experiment, that you are interested in ten or twenty such projects in a variety of organisations. You have no time for all the detail, but want to be involved in the *major decisions* taken about each project. Moreover, you want to ensure that the people on each project do recognise the issues for decision and resolve them consciously, instead of just letting things happen by default.

Certain generic issues for decision will arise in project after project. For example, how much effort should be devoted to modelling the present workings of the business process that is to be re-engineered? This question arises on nearly every project, and different answers are sensible in different situations. It is a shame for a re-engineering team to drift into modelling at a certain level of detail as prescribed by some standard methodology, without recognising that degree of modelling detail is a variable for decision on each project.

There are perhaps several dozen generic issues for decision that can have a large effect on the outcome of a re-engineering project. This briefing advances the claim that all these issues can be grouped under four headings, or better, that there are *four main kinds of decision* about re-engineering. Once accepted, this concept can be a powerful tool for sorting out more detailed material, such as theories, methods, examples, points of controversy, and so on.

▼ Going for the decisions is a good general-purpose technique for cutting through *any* thicket of theory about management or about IT. While reading a whole book about (say) end-user computing or quality management or corporate excellence, get things into focus by asking: What are the main *decisions* that organisations typically need to take about this subject? Confronted by a prolix case-study or a turgid text about megatrends in society, ask the question: What *decisions* could this material help anybody in an actual organisation to take, and how? The better the answer that is forthcoming, the better the quality of the ideas, and vice versa. Abstract academic theories or gushing anecdotes about innovative companies that have no clear relevance to any practical decisions about anything tangible can be consigned to the flames. ▲

REPRESENTATIVE IDEAS

The briefings in this book are structured to bring together in one section a selection of ideas from books and articles that are either influential or stimulating or both. However, none of the items listed in the bibliography breaks down the generic decisions in re-engineering. Therefore, in this briefing, the section is void.

DISCUSSION

The term *process* is customarily used for the chunk of an organisation's operations that is re-engineered. Decisions about the re-engineering of a process can be partitioned into four kinds.

Distinguishing Four Kinds of Decision

Two kinds of decision are associated with the *setting up* of the project to re-engineer a process; a third kind of decision is about the *characteristics* of the process design that emerges; the fourth kind is about the *implementation* of the new process in the organisation.

● **Project scope decisions.** Plainly, some early decisions must be taken about the terms of reference to be given to a re-engineering design team: at its simplest, one particular process rather than others is chosen to be re-engineered. In some situations, the right scope decision may be obvious and one meeting will suffice to establish it. At the other extreme, several months' work investigating the business's strengths, weaknesses and opportunities may precede the decision to establish a re-engineering project with a certain scope.

● **Design approach decisions.** Once a scope decision has been taken, and before detailed design work gets under way, decisions are needed about the way the design work is to be done. Some of these will shape the whole character of the work: eg 'Should we take a strongly quantitative approach, making simulation models to study cycle-times under different volumes of transactions — or is that inappropriate for this particular project?'

● **Process design decisions**. The product of the work of re-engineering design is a new design — perhaps quite a detailed document. Since there may be thousands of details and most could conceivably have been otherwise, this design embodies a great many design decisions. But probably only a small minority can be regarded as major decisions, choosing between significant options: eg 'Shall we adopt *this* extremely radical design for a new process or *that* one, only moderately radical, but perhaps more prudent?' It is true that some re-engineering teams make a design without explicitly choosing between major options. But in practically any such situation it is possible to point out some more elaborate option and some more modest option, neither of which are nonsensical; thus, even if not documented or even noticed, design options are being rejected implicitly. Decisions of process

Re-engineering: Kinds of Work and Kinds of Decisions

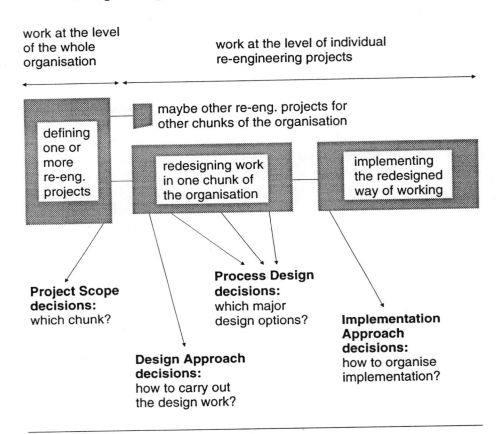

work at the level
of the whole
organisation

work at the level of individual
re-engineering projects

maybe other re-eng. projects for
other chunks of the organisation

defining
one or
more
re-eng.
projects

redesigning work
in one chunk of
the organisation

implementing
the redesigned
way of working

**Project Scope
decisions:**
which chunk?

**Process Design
decisions:**
which major
design options?

**Implementation
Approach
decisions:**
how to organise
implementation?

**Design Approach
decisions:**
how to carry out
the design work?

design can occur at various points in the design activity: near the end, if the team works out two or three possible designs in detail to be compared for final choice; or much earlier, if the team recognises some crucial choices of principle to be made before detailed work is done.

● **Implementation approach decisions.** To design is one thing, to implement another. There is rarely just one self-evident way of phasing-in and introducing the new process. Thus decisions about the approach to implementation belong naturally towards the end of the design work or soon after.

The diagram gives an impression of the way these four kinds of decision are related. As it shows, decisions of project scope are based on work done at the level of the whole organisation (or at any rate a wide part of it), since they entail choices in favour of re-engineering certain chunks at the expense of others. The other three kinds of decision are taken within one particular re-engineering project.

The lists over the page show some arbitrary, representative examples of the four kinds of decision.

▼ This *analysis* is a different thing from a *recommendation* that re-engineering work should always have some particular structure. It simply points out that any re-engineering undertaken in a reasonably competent way will call for certain decisions, and, however the work may be organised in detail, the decisions will be of the four kinds described.

Many consultants have a standard multi-stage approach to offer. One common idea is that you should, first, establish the scope of a project; second, study the current process; and third, design the new process. The analysis of kinds of decision in this briefing doesn't imply that this is either good or a bad. It asserts that, whether you work in three, six or 27 stages, you will need to make these *four kinds of decision*. How to divide a particular project up into activities is itself a decision of the design-approach kind. ▲

Change-management and Directly-rational Factors

Decisions of these four kinds will result from weighing up certain factors in what may be called a *directly-rational* way. If the scope decision is taken to re-engineer one part of the business rather than another, there are presumably some cogent arguments, such as 'Our process A is vastly inferior to the equivalent process at our competitors' or 'The chance of gaining spectacular benefits is greater with process A than with B.'

Decisions about Re-engineering

Kind of Decision	Representative Example Decision
Project Scope	'Our business divides neatly into eight obvious chunks; we will set up two re-engineering projects: one for A and one for E; and we won't re-engineer the other six.' *As opposed to* 'Out of the eight chunks we will initiate re-engineering for A, D and H.' 'The most natural way of analysing our business shows 20 main chunks; we will set up one re-engineering project to tackle the whole of L, M and N taken together as one unified problem.' *As opposed to* 'L, M and N will each have its own re-engineering project.' 'We have mapped out our business as 15 continent-like chunks, but the area we will re-engineer as one unified project can't be fitted simply onto that map; it contains most of chunk P, about half of Q, and a small part of R.' *As opposed to* 'We have analysed our business into 15 possible, coherent chunks for re-engineering. We will re-engineer the whole of S in one project, and the whole of T in another, separate project.'
Design Approach	'We will start by making a fully detailed model of the way the current process works — perhaps 50 pages long — containing much quantitative data about volumes, times and costs.' *As opposed to* 'We will document the existing process only roughly — perhaps in five pages of diagrams without any quantitative content.'

'We will design a new process in fair detail first; after that, we will check that its IT implications are acceptable.'

> *As opposed to* 'A major factor in our new design will be recent developments in information technology that hardly any other business has yet used. Our design work will be organised to stimulate bright ideas of that kind.'

'Most of the design team's work will be organised around a prototype version of a new process, that will be constantly modified to try out a variety of design possibilities in a fairly unstructured way.'

> *As opposed to* 'The design work will be methodically structured into four successive stages of increasing levels of detail.'

Process Design

'Three possible process designs seem plausible — A, B and C. Each has advantages and disadvantages. None is superior on all criteria: strict cost-benefits, risk, implementation time, service quality, job enrichment etc. After weighing things up, we choose A.'

> *As opposed to* 'On balance, we choose B.'

'All the main factors associated with the four main design options — D, E, F and G — can be translated into financial terms. Thus the four options can be compared on DCF principles. We choose the one that comes top on this basis — D.'

> *As opposed to* 'We choose F, which calls for half the investment of D, is far less risky and still offers an attractive return.'

'We identify one basic improved process, together with five optional extra features, H to L. We choose the basic version together with features H, I and J; we discard K and L.'

> *As opposed to* 'We choose the basic version of the process. Maybe, though we make no commitment, we will add on some or all of the extra features at some later point.'

Implement-ation Approach	'The newly designed process can't be introduced instantaneously throughout the company, but it will be phased in branch by branch as quickly as it reasonably can.'

As opposed to 'The complete newly designed process will go in as a pilot at one regional office. After a six-month period we will probably make some third- and fourth-order changes of detail. Only then will we decide on how to phase it in to the whole company.'

'The newly designed process can be split into three segments. First, we will phase in segment one throughout the company; after that will we proceed with segments two and three.'

As opposed to 'The new design can't reasonably be split into segments. We will introduce the complete process at one regional office after another.'

'Separate teams will be responsible for re-engineering design work and for implementation of the process in the branches. There will be feedback between the two teams.'

As opposed to 'The same team will handle both design and implementation.'

But there is a complication. Many, many authorities point out that *change management* is a very big part of re-engineering. In this context change management means something far more specific than just 'anything associated with managing change'; it stands for the management of the emotional, non-rational aspects of change.

Even if process A seems a better choice for re-engineering than B, on directly rational, analytical grounds such as those given above, there may still be contrary, psychology-dominated, change-management factors: eg 'Everybody in the organisation knows that whenever a new management fad or a new piece of information technology is tested out, A is the guinea-pig process.

But we need to show our staff that re-engineering is something special, not just the flavour of the month. Therefore we should avoid A, and choose B instead. People will realise that, since B is such a vital part of the business, we would never choose it for re-engineering unless we were deadly serious.'

This argument may or may not prevail; it depends on the strength of other pertinent factors in the specific situation. The point is that in many decisions about re-engineering there are two types of factor to be considered: directly-rational factors and change-management factors.

Now a complication arises. In practice, the great majority of insights about re-engineering worth reading about concern directly-rational factors. Much advice about change management in books, articles or seminars is nothing but exhortation to recognise the obvious: remember that some people are instinctively afraid of change, avoid humiliating people, demonstrate top-management commitment etc. In any book that avoids lingering over the obvious, the proportion of chapters and pages specifically devoted to change-management factors can't correspond to the weight these factors have in some actual projects. There is no escaping this imbalance.

▼ Important but analysis-resistant topics are found in many fields. A textbook of accountancy may contain a great deal about possible methods of valuing inventory, but hardly any exhortations to the accountant to be honest. And yet for a successful career in accountancy, it is probably better to be honest but ignorant of the finer points of valuing inventory than to be an expert on inventory valuation with several convictions for theft. Not stealing from the petty cash is very important indeed, and so is not humiliating the people who will use the newly re-engineered process, but on both subjects, there is little more of a general character that is worth reading about in a book. ▲

The Briefings in This Book

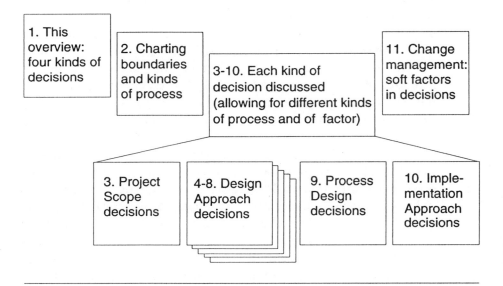

CONNECTIONS

The heart of this book — Briefings 3-10 — is an account of the various issues that typically arise with the four kinds of decisions. As the diagram shows, these briefings are framed by two others.

Before investigating particular decisions it is best to set boundaries to the kind of project to be counted as re-engineering. Also, if projects within the re-engineering category can then be broken down into several distinct varieties, many issues will be dissected with a sharper knife. The charting of boundaries and breakdowns is done in Briefing 2.

Change-management factors crop up along the way through the main body of briefings, but Briefing 11 deals specifically with some puzzles of change-management practice.

2.   Boundaries and Breakdowns

Certain case-studies are universally reckoned to be classics of re-engineering. But not all cases can be classic. What determines whether some particular change initiative ought be counted as re-engineering or not? This question is closely related to another: Once the field of re-engineering has been demarcated, can it be sub-divided into several distinct territories?

Boundaries

Most books on the subject define re-engineering, using words such as 'radical' or 'dramatic'. But *how* radical must change be to count as re-engineering? There is a more interesting question too: Does any change to any aspect of an organisation count as re-engineering, if it is radical enough, or can there be some radical changes that are not re-engineering?

Nobody is likely to deny the term re-engineering to famous examples such as Ford reorganising its materials procurement procedures or IBM Credit finding better ways to process applications for financing computers. But suppose a publishing company reshapes itself by selling off one division, reshuffling three other divisions into two, and generally shifting the emphasis from books to magazines, and from the academic to the professional market? That is radical change, but rather different from the classic examples of Ford and IBM Credit. Its instigators may well claim to be re-engineering their business. Is that sensible?

What about Taco Bell, one of Hammer and Champy's examples? It installed new kitchen machinery, altered the physical layout of its restaurants and reorganised the responsibilities of its managers. That also seems rather different from the classic examples of re-engineering.

Breakdown

Is it advantageous to make a breakdown *within* the whole class of re-engineering projects? Certainly there seems to be some variety of traits. With Ford and IBM Credit the redesigned process is a crisper, more streamlined machine for handling large quantities of transactions efficiently. There may be multiple paths for different kinds of transaction, but still the process possesses a definable pattern of operations. Many of the best-known case-studies are like this, but some are rather different.

One of Hammer and Champy's main examples is Kodak's process for developing a new camera model. This process is nothing like an efficient transaction-processing machine; it enables a variety of people — camera body designers in one office, shutter designers in another, and so on — to work together in parallel and interact in complex, unpredictable ways.

This suggests that at least two kinds of process may be re-engineered — and, on more careful analysis, very likely more. There is an attractive prize to be sought here. A framework distinguishing several kinds of process for re-engineering, if sound, can be a tool for probing specific issues: process modelling, simulation, the role of IT, change management etc. On each issue in turn, it can promote reasoning along the lines: for a type-a process the main options to consider on this issue are . . . but for a type-b process . . . This should make it easier to arrive at perceptive decisions based on situation-specific factors.

REPRESENTATIVE IDEAS

There is little point in marking out boundaries to the field of re-engineering greatly different from those commonly understood. A better aim is to get a broad impression of the way the term is typically used, and then decide how to tidy certain frontier areas that may otherwise cause confusion. There are three main angles: definition of re-engineering in *abstract words*; *typical features* (saving money, speeding up procedures etc) of well-known examples; and the *nature* of the examples (type of work, scale of change etc).

Formal Definitions

Hammer and Champy provide a paragraph-length definition that talks of redesign leading to improved business performance; they say their definition has four key words: fundamental, radical, dramatic and process.

There is redundancy here. Practically any change that is *fundamental* is pretty sure to be *radical* as well; moreover, *dramatically* successful changes are most likely to be the product of radical and fundamental, rather than timid and shallow, design.

Hammer and Champy talk vaguely of a *process* consisting of activities that take inputs and create outputs of value to the customer. Taken as a whole their book gives an impression of a process as a chunk of the organisation's activities, that: a, is substantial; b, can be isolated fairly cleanly from everything else; and c, probably crosses boundaries within the hierarchy of the organisation.

Most others define re-engineering in some combination of abstract words that amounts to much the same. Morris and Brandon are an eccentric exception. They seem to wish to call *any* change to a company's procedures — however large or tiny —

re-engineering, provided only that it occurs within the framework of their own methodology for planning and organising change.

▼ Warning: consultants hunting for work are often guilty of this kind of thing. It is much easier to fill up glossy brochures with long lists of successfully completed assignments in re-engineering or total quality management or concurrent engineering or anything else of that sort, if the term is treated as infinitely elastic. ▲

Typical Features of Re-engineering Examples

A more fruitful line is to pick out distinctive features of the re-engineering classics. The original Hammer article describes two main examples:

● **Ford** re-engineered its purchasing, goods inwards and accounts payable tasks. Formerly, numerous documents such as purchase orders, invoices, delivery notes and so on flowed between these three departments and Ford's suppliers. Much effort was spent on sorting out inconsistencies between documents or between documents and delivered goods. The new process makes more use of computers and includes some radical innovations. If a delivery of goods doesn't correspond to a purchase order stored in the computer system the goods are refused; there are no messy procedures to resolve discrepancies. Suppliers are told not to send invoices (any that come are ignored); payments are simply sent out automatically for all goods accepted. The spectacular result is that Ford's accounts payable department, formerly containing 500 people, has been reduced to 125.

● **Mutual Benefit Life Insurance** re-engineered the process of handling applications for new policies. There used to be 30 procedural steps, five departments involved and 19 people. Now one case manager is responsible for most of this work. He or she has an expert system on a PC connected to a mainframe computer. Turnround is much faster, and the same number of people can handle twice as many new policies as previously.

Most of the detailed examples in the Hammer and Champy book contain similar themes: IBM Credit streamlines its processing of requests from customers for loans and leases; Bell Atlantic redesigns its administrative procedures for connecting new customers; Capital Holding makes its direct marketing of insurance more efficient; and Imperial Insurance (an imaginary but rich example) rethinks its processing of claims. Three notable aspects of re-engineering seem to recur:

● Successful re-engineering produces spectacular **quantitative benefit** (eg 125 people do the work of 500) — as opposed to benefits that are merely healthy (500 down to 400, say).

● As the Mutual Benefit Life Insurance example shows, **job enrichment** may follow too: workers can perform a richer mix of activities.

● The third common aspect of re-engineering seems to be **simplicity**, or perhaps elegance. A chart representing the procedures and associated documentation of the new process will be much simpler.

Nature of the Examples

Here are two strong impressions that arise from studying the nature of the examples found in the most prominent sources:

● It doesn't sound glamorous but the great majority re-engineer **office work** — provided that term is taken broadly enough to include the office procedures within (say) a factory's goods inwards department. Ford may perhaps make radical improvements in the factory by using sophisticated robots, but that kind of thing is hardly ever cited as an example of re-engineering. Davenport and Short, in their article, put the following argument. In recent decades, through IT and other factors, companies have made far greater improvements in shop-floor work than in office work; and what improvements there have been in office work have often been related to the shop-floor, eg production scheduling or material requirements planning. Therefore the striking opportunities for improvement *now* are in service-related office work. This theme is reiterated, though much less forcefully in Davenport's book. (Several other vivid points from the original article lose impact in the drab Harvardspeak of the book.)

- The stress is usually on **process innovation** (finding better ways of providing much the same product or service — though often with second-order changes and extensions) rather than *product innovation* (finding completely new products or services). Macro-level changes in an organisation, such as a publisher reshuffling its main divisions and changing its market stress, are not usually called re-engineering — except very loosely. Thus Mutual Benefit Life Insurance may now offer improved service to policy-holders, but its re-engineering is nowhere near as radical as entering the market to insure ships and aeroplanes.

Hammer and Champy don't make either of the above points explicitly, but most of their examples bear witness to them:

- The enthusiastic account of **Hallmark**, makers of greeting cards, may suggest that radical changes are being made to everything conceivable. But, read carefully, it contains no mention of changing the manufacture of the physical products, the cards. Printing is scheduled better but there is nothing about new printing machines or redesigned layouts for the printing plant and its stores.
- **Bell Atlantic** finds better methods for scheduling the installation of new telecoms equipment, but it doesn't invent new machines for actually doing the installation work.
- **Imperial Insurance** considers visiting accident victims in hospital, even those insured with other companies — not only to gain goodwill, but to see that the person gets 'good but not unnecessary' medical care. The team also considers taking the initiative to *arrange* for an insured to have a hired car as a temporary replacement, as opposed to just *allowing* the insured to make the hire and claim the cost afterwards. However, it doesn't consider first-order changes of business, such as taking over a chain of hospitals or a car-hire firm.
- Changes of demarcation, such as shifting work and responsibility across to suppliers, are common in re-engineering examples. For example, **Ford** picks one favourite supplier for brakes and sets up closely interlocked administrative arrangements with it. This is quite a change, but it is still of a lesser order

than Ford building a new factory of its own to make brakes, and then marketing them to the whole industry.

▼ To make such points is not to criticise or praise. To think clearly about *any* class of things — re-engineering projects, gemstones, tramways, post-modern architecture or anything else — you need to be clear about the main traits of things included in the class. Otherwise, you can't proceed confidently with inquiry into the subject. ▲

Some Exception Examples

Hammer and Champy have one detailed example that, though they don't remark on it, is quite different from all the others. Scrutinised carefully, the main changes made by the successful fast-food chain Taco Bell are these six:

● The company widens its target market very ambitiously: from regional to countrywide; from Mexican specialities to other fast food; and from just high-street restaurants to other outlets too (eg schools, airports etc).

● Expenditure on food is increased, and on marketing correspondingly reduced.

● The old bureaucratic management structure is slimmed down.

● Restaurants are completely redesigned. The average ratio of kitchen to customer area was 70:30; now it is 30:70.

● The previous point is made possible because work in the kitchens is greatly reduced. Meat and beans and other components of the meal are cooked at centralised locations, and just reheated at the restaurant kitchen.

● New technology for fast-food mass-production is installed. There is a new taco-making machine: 900 perfectly proportioned tacos an hour, and ready wrapped too.

Most of these points are either macro-changes of business policy, or else changes to physical operations rather than office work.

Carr et al, among perhaps 20 or 30 brief examples designated as re-engineering, include two that lack the usual traits:

● A **drug company** used a supercomputer to model organic molecules in order to discover new drugs more effectively.

● The *Farm Journal* developed systems to produce 5000 different versions of each edition, with customised selection of editorial material and adverts for distribution to 800,000 farms. This required a database of detailed information about each subscriber; the database facilitated other lines of business, such as conducting market research for makers of farm products.

DISCUSSION

The abstract definitions mentioned above lead nowhere very fertile. The way to find good boundaries for re-engineering is to study the representative examples.

Three Recurring Features

On a casual reading of many examples the three characteristic features of spectacular quantitative benefit, job enrichment and simplicity seem to recur endlessly. But, on closer examination, these three don't always go together. In the Ford example spectacular quantitative benefit and simplicity are very pronounced, but there is hardly any stress on job enrichment. In the Mutual Benefit example most of the stress is on job enrichment and simplicity; quantitative benefit is impressive rather than spectacular.

To count as *classic* re-engineering, it might perhaps be insisted, an example should possess all three of these features to a substantial degree. Since this is quite a stringent test, it seems sensible to allow for non-classic examples too. Spectacular quantitative benefit might be aimed for and achieved, without making jobs richer or poorer, for instance. Or, though few examples are described explicitly like this, an organisation might give many people wider, richer jobs — and justify this by long-term, unquantifiable considerations, rather than by spectacular immediate benefits. Again, a new process might bring high quantitative benefit (eg faster throughput), by becoming more, not less, complex than the old. Davenport's book cites Phoenix Mutual Insurance and Federal Mogul (makers of car components), where the new processes that cut cycle-time were more complicated than the old.

▼ When anything appears (or is said) to have three salient features, it is invariably worth asking, as an almost automatic reaction: Are these three always found clustered together, or not? This often yields practical benefit. Being aware of three classic, but not essential, traits of re-engineering makes a good start towards exposing potentially valuable *design options* in a particular situation, each giving the factors different stress. Awareness of more options should, in turn, lead to better decisions. The technique of starting from the classic traits and varying them to generate options is applicable to decisions in many other fields besides re-engineering. ▲

Charting Boundaries

To settle some boundaries between re-engineering and other things begin by distinguishing two activities:
- Taking **strategic business decisions**; eg a publisher reorganises three divisions as two, shifts market focus from educational to professional, etc.
- Conceiving and implementing **new process designs**; eg the job content of an acquisitions editor is changed, the elapsed time in publishing a book is reduced, etc.

It seems best to keep the term *re-engineering* for the second of these. True, you may do both: adopt a new organisation and market strategy, and also redesign processes for the new organisation. But you may also change business strategy without radical changes to processes. Or you may redesign processes, without any fundamental change of business strategy.

Here are some further defining characteristics that seem worth stipulating for a project to count as re-engineering:
- The project aims to make **radical changes**.

● The project is directed at **substantial improvement** in at least one of the following three factors: quantitative benefit (not necessarily directly financial, perhaps order fulfilment time or customer reorder rate); job enrichment; and simplicity.

● The process addressed handles **office work**. Thus changes to physical systems, such as inventing and introducing advanced taco-making machines, are excluded.

These boundaries enclose the majority of re-engineering examples actually cited, and provide clear grounds for excluding the minority that do indeed seem rather separate (eg a farm magazine developing entirely new types of product, or a drug company modelling organic molecules). Thus they demarcate a coherent field for further inquiry.

The inclusion of 'changes' in the suggested boundary-drawing may suggest that the design of processes for a completely new enterprise can't be re-engineering, since there is nothing pre-existing to be changed. But examples such as GM setting up sophisticated systems for a new luxury-car division are sometimes cited as re-engineering. This is of no great concern, since the designers start out with an awareness of the way car production is usually managed and their aim is to improve on that.

Davenport, in his interview, notes perceptively that sometimes the best buy (ie best balance of advantage relative to cost — financial, psychic and other) may be to leave the main features of a certain process untouched, but alter one or two other variables such as allocation of responsibilities, method of measuring performance, or method of employee compensation. He gives the example of Xerox gaining 50% improvement in supply chain management, by altering managers' responsibilities and compensation without any redesign at all. For him, such changes are not re-engineering, and they are indeed excluded by the boundaries suggested above.

▼ Avoid the fallacy of expansive categories. To exclude Seurat from the category of the Impressionists is not to say that his work is unimportant. The aim is to make the category of Impressionist as useful as possible; Seurat might be your favourite painter, but you could still feel that calling him an Impressionist made that category too broad to be coherent. Similarly, to exclude taco-making machines and supercomputers that model organic molecules from the category of re-engineering is not to denigrate those innovations. It is merely to suggest that if they and all similar examples are counted as re-engineering, the category will be so widely drawn as to impair well-focused analysis. ▲

REPRESENTATIVE IDEAS

Now for the second large task: establishing categories within re-engineering. There are two main ways of doing this: by the *nature of the business process* subjected to re-engineering, and by the *business effects* of the re-engineering project.

Kinds of Process: a Complicated Breakdown

One complicated way of categorising business processes that may be re-engineered is to assess any one process on three dimensions:
● The process may be interfunctional (ie across several departments of the organisation); or interorganisational (ie involving business partners, such as customers or suppliers); or interpersonal (within one department).
● The process may be physical (eg manufacturing) or informational (most office work).
● The process may be operational (day-to-day) or managerial (control, planning etc).

Since any process can have any combination of these attributes there must be 12 (3x2x2) kinds of process. This suggestion comes from the article by Davenport and Short. It doesn't illustrate how to locate any given process along each of the three dimensions. Are all twelve combinations really possible? Can there be a process that is within one department, physical and managerial, for example? The quibbles flood in:
● Surely not every process is *intrinsically* interfunctional or interorganisational or interpersonal. A process might be re-engineered in such a way that boundaries were redrawn. Allowing

for this possibility would make the whole framework even more complex.

● The article distinguishes between physical and informational processes, but then confuses everything by saying that in many physical processes adding information to the object adds value, eg as in Federal Express's parcel tracking service.

● If every process is either operational or managerial, where does this leave product development (eg Kodak developing a new camera)?

This breakdown is best regarded as a stimulus to further thought rather than something usable as it stands. Why mention it then? Because of the surprising fact that few people other than Davenport have worked out and published any breakdown of kinds of process.

Kinds of Process: a Simpler Breakdown

Davenport's book abandons the breakdown just given in favour of a much simpler one:

● First, make a distinction between processes that have an entirely **physical** output (eg manufacturing a paperclip or giving a haircut), and those that have an entirely or partly informational product — in practice, virtually all the processes that come up in debate about re-engineering.

● The processes with **informational** products can be divided three ways between: management processes; operational processes that are transaction-driven; and operational processes that are unstructured.

● An example of a **management** process is IBM's system for analysing market and economic data for each of the countries of Latin America.

● An example of a **transaction-driven operational** process is the handling of cheques in a bank or, in fact, almost any of the best-known re-engineering examples.

● **Unstructured operational** processes are, to be more precise, *relatively* unstructured processes that could with advantage become more structured. Examples are: an IBM system for market information (containing not just sales analysis, but also customer feedback, intelligence about competitors etc), and a consulting firm's skills database (containing consultants' CVs, texts of their external publications, texts and graphics of consultancy reports and presentations etc).

Davenport says more about management processes than any other author — but only to recommend against re-engineering them at all, since they are so resistant to structuring. He advocates 'gentle movement' towards making things rather more firmly structured, rather than brusque introduction of radical new processes.

A large drawback with his breakdown is that the distinction between the 'unstructured operational' process (eg IBM's system for market information) and the 'management' process (eg IBM's system for Latin America information) is obscure. And if those two kinds are merged, the analysis collapses into transaction-oriented (the overwhelming bulk of re-engineering examples) and everything else (mostly not to be re-engineered).

Elsewhere in the book other categorisations are given, but they are far more literal in character. For example:
● Order management is placed as one generic process, within the broader category of marketing processes, within the still broader category of customer-facing processes.
● Management processes can be analysed into the following seven: strategic decisions; planning and budgeting; performance measurement; resource allocation; human resource management; stakeholder communications; infrastructure-building.

These detailed, literal breakdowns are just a convenient device for arranging ideas and examples in an orderly way within a book; no more than this is claimed.

▼ To say: 'These seven headings help sort out most of the examples of an X (a management process, a marketing policy, an investment opportunity etc) that we have found' is not the same as saying 'Every example of an X belongs to one and only one out of these seven (not six or eight) categories.' Confusion and fallacy can arise if somebody starts out asserting the first, and continues by reasoning as if the second, much stronger, claim has been accepted. ▲

'Product Development' Processes

Davenport's work does at least suggest that there can be kinds of process other than the administrative ones of Ford or IBM Credit. Other authors are not so explicit about this point, but a number of case-studies demonstrate it. Quite a few processes can be provisionally labelled as *product development:*

● **Kodak**. Hammer and Champy describe a new process for developing one new camera product; the elapsed time (38 weeks) was half that usually expected. The essential difficulty with this kind of task is that purely sequential work (develop the camera-body first; freeze that design; then start shutter design . . .) is far too slow, while parallel development may lead to anarchy, since the parts designed separately may not fit together. Any approach has to be some compromise between the two. A re-engineered process making good use of CAD/CAM database technology and telecoms links between separate design teams can shift the balance of tradeoffs, and allow much more to be done in parallel without anarchy.

● **Hallmark**. The same problem exists with greeting cards. To an outsider it might seem easy to co-ordinate artists drawing Christmas scenes and poets writing festive messages, but apparently it is not. In Hammer and Champy's book the Hallmark people tell how, after radical re-engineering of this process, a new Christmas card can be produced in less than a year's elapsed time; that may not sound brilliant, but at Hallmark they are delighted. Here there was no bold use of technology; the key reforms were integrated teams (artists and poets working together in the same office), and reduction of the power of outside managers to keep altering the product.

● **Bendix**. Carr et al give the example of the development of new brakes products for motor vehicles. It seems essentially similar to the Kodak example.

● **McDonnell Douglas.** In this Kodak-like example, from Morris and Brandon, the product designed is an aircraft. Moreover, the re-engineered process encompasses not just the production of design drawings, but also the technical documentation delivered to the maintenance engineers of the purchasing airline.

● **AT&T.** Johansson et al describe the design process for an OLS product (some kind of electrical device). Each product is different and has to be first ordered by the customer, then designed and then manufactured. However, this is a long way from the Kodak example. The designers choose which out of numerous components, sub-assemblies and standard designs to combine together and perhaps customise. Re-engineering cut the average design and production time from 53 to 5 days, mainly in these three ways: rationalisation of procedures through common sense; bringing designers of different disciplines together in teams at the same location; and having more standard building-block combinations of components ready-made in stock (even at extra inventory cost and some loss in optimality of design).

'Managerial' Processes

Another broad group of processes, that are neither routine administration nor product development, can be provisionally labelled as *managerial*:

● **Hallmark.** Part of Hammer and Champy's Hallmark case-study is the management decision-making process. Point-of-sale devices read barcodes and send sales data (ie individual purchases by consumers) from stores straight back to Hallmark. This feeds new decision-support systems, where sales trends are represented graphically, sales forecasts are generated, and assessments are made of the effectiveness of a store layout or of an advertising campaign.

● **Bow Valley.** The *Economist* article of May 1993 gives the example of an energy company that introduced new systems, based on high-powered desktop workstations, to share management information between divisions. This permitted three layers of management to be stripped out.

● **Xerox**. Davenport and Short describe an executive information system for consolidating and reviewing the plans of each division. Information is sent over a telecoms network to senior managers, who not only access but also amend data through workstations during meetings.

● **Texas Instruments.** Davenport and Short also mention an expert system for the capital budget request process — reducing preparation time from nine hours to 40 minutes, and bringing better compliance with company standards.

Other Interesting Processes

A few of the case-studies to be found in the literature don't fall under any of the headings suggested so far, but seem worth noticing:

● **Rheden.** Dur's book describes the process of handling requests for building-permits by the authorities in a small town. About 600 requests are received per year. The essence of the processing is to check the request against a variety of criteria; thus, there is some similarity to the Texas Instruments capital-budget example. But, however complicated the criteria-checking may be, there is another fascinating feature too: part of the work is to serve the customer by explaining why a request falls short, and giving advice on how it might be amended to have a better chance.

● **Telstra.** Bendall-Harris describes how Telstra, the Australian telecoms utility, re-engineered its procedures for handling customer complaints. It seems incredible but throughout the company there were 50 different complaints procedures to be rationalised. This example records no dramatic statistics of quantitative improvement, but it raises some good questions: among others, how do you measure success with this kind of process?

● **Ontario PM.** Proc et al describe re-engineering of the process of handling letters from the public in the office of the prime minister of Ontario. The volume of letters rose from 137,000 in 1991 to 312,000 in 1993; but during this period average response time was brought down from 44 days to 12, and staff was reduced from 42 to 32. The aim with this process is to deal with letters efficiently, but without seeming perfunctory, and also (this is the tricky part) with adequate screening and reviewing to avoid sending a reply that might be regretted later. The old process flow, if fully charted out, had about 300 separate steps, the new one only 30.

Breakdown by Status of Process

The examples just given are sorted roughly by the kind of thing that the process does. Edwards and Peppard suggest a different angle that might be called the *status* of the process:

● A **competitive** process is directly related to the current chosen way of competing with other firms; eg order fulfilment, if (and only if) the firm regards efficiency in fulfilling orders as a key issue in competing against rivals.

● A **core** process is necessary to the business but is not a key competitive issue; eg a vehicle-scheduling process may be essential to a distribution business, but not the basis of competition with rivals.

● An **infrastructure** process creates the capability to compete in the future: this seems to apply mainly to training and R&D activities.

● An **underpinning** process exists but is barely recognised as a discrete process; eg common administrative support for several of the other processes: no example is given, but a process like payroll may be intended.

Sceptical questions flood in, but the main one is this: What has this to do with re-engineering? The authors claim that there is some connection, but is there? If you classify your organisation's processes in this way, what re-engineering decisions are you better able to take? Choice of process to re-engineer? Method of modelling the process? Change-management style? Role of IT? If not these, then what decisions? The article does not say.

▼ Ask of any proposed categorisation — of processes, of competitive forces, of canteen-management paradigms, or of anything else: What decisions would this help you with, and how? This is a swift test of the quality of any management thought. ▲

Breakdown by Business Effects

Another concept is to break down re-engineering projects by the business effects that flow from the new version of the process. Here, summarised from Johansson et al, is the logic of one such breakdown.

Any successful re-engineering project brings an improvement to some feature of the business: cycle-time for orders, clerical costs etc. But, though great improvement *may* have great consequences, such as capturing a much bigger market share, it may not. A successful re-engineering project may have one out of three possible *business effects*:

● **Breakpoint**: much bigger market share, or some comparable benefit that makes the business outstanding in its industry. A Coca-Cola and Schweppes joint-venture bottling-plant was re-engineered to achieve three main benefits: drinks bottled much quicker; peak summer demand met by improved utilisation of capacity; cost of making and storing cans reduced.

● **Parity**: benefits of great importance — not because they achieve a dramatic advantage in the market, but (usually) because they are essential to keep up with the leading pack in the industry. As described earlier, AT&T designed its made-to-order OLS products much more rapidly.

● **Improvement (only)**: considerable benefits, well worth having, but of second-order importance, since they don't affect the business's position in the market very much. Dun and Bradstreet, supplier of financial information, cut its cycle-time for processing contracts with new customers.

These three kinds of *business effect* are not identical to the measured *quantitative effect* of a project. Coca-Cola and Schweppes achieve (only) a 50% cut in cycle-time for bottling; AT&T cuts design and production time by 90%, from 53 days to 5; and Dun and Bradstreet cuts its time to process new contracts from more than one week to less than one day. The logic here is that in the soft-drinks business (presumably) a 50% cut in cycle-time for bottling brought great business advantage, whereas, for Dun and Bradstreet, even a huge cut in cycle-time for new contracts had little business effect. In the AT&T example, according to the book, a cut in cycle-time (however great) could not be the factor that swept the market; rather it was a compensation for certain unassailable cost advantages enjoyed by competitors.

There is no implication that the three gradations represent a measure of project success. An improvement (only) project isn't (or isn't necessarily) a failed breakpoint project.

Breakdown by Effect on the Industry

Johansson et al sow confusion with talk of achieving market dominance by changing the rules of the industry. In fact, their breakpoint category is broad enough to include *any* improvement that poses a severe threat to competitors, including such a mundane thing as delivering the same soft drinks in much the same way, but faster, more reliably and cheaper. Most other writers reserve the language of changing the rules or altering the structure of an industry for moves to offer some entirely new product or service; eg a new service to manage customers' inventories of canned drinks, allowing for weather forecasts, holidays, promotion campaigns etc.

An influential but unduly prolix article by Short and Venkatraman describes the example of Baxter, a supplier of goods to hospitals. Over the years it has provided an ever-improving service for customers to order goods. At a certain point it took the large step of introducing a 'materials management' service. If a hospital signs up for the service, Baxter undertakes to monitor usage of supplies by individual departments within the hospital (operating theatres, X-ray units etc), and to deliver replenishments on a just-in-time, seven-days-a-week basis, direct to each department. This can reasonably be called an attempt to alter the structure of the medical-supplies business.

Thus these authors, and some others too, make a distinction between:

• re-engineering to do the things currently done much better; eg collecting and meeting customer orders more efficiently;

● re-engineering to do things arising from a new definition of what is to be done; eg offering a 'materials management' service, that in effect redefines the whole notion of a customer order.

In his defensive *Economist* article of late-1994 Hammer offers examples that illustrate, albeit only implicitly, a similar but different two-way distinction:

● re-engineering to do the things currently done much better; eg Texas Instruments cuts order fulfilment time by 50%;

● re-engineering to do things not currently done, though already done by others, ie to break into new markets; eg Progressive, a motor insurer, branches out into insuring new kinds of risk, challenging established rivals by offering a superior, re-engineered process for handling claims.

DISCUSSION

The main criticism of the breakdowns by kind of process contained in Davenport's book is that they don't lead to any decision-assisting insights such as that one kind of process is best approached in a certain way, whereas for another kind another approach is more fitting. This section works out a framework to serve that purpose.

Distinguishing Four Kinds of Process

For most re-engineered processes there is an obvious *unit of process*. At IBM Credit, the unit is one application for credit; at Ford the unit is an order for a certain part on a supplier; at Mutual Benefit, an application for a new life assurance policy. In other words, any chart describing the process is essentially an account of the life — or rather, range of possible lives — of one credit application, order or new policy.

Most of the famous examples are of processes that handle large quantities of units in a pretty standardised way. That is why statistics about great reductions in average cycle-time are meaningful and impressive. But for Kodak's product-development process the only meaningful unit is the design of one entire new camera model. Though IBM Credit may handle hundreds of credit authorisations a week, and Ford hundreds of part orders a day, Kodak doesn't design hundreds of new camera models even in one year. This unit of process, a complete design for one new model, is a much more substantial thing — measured by quantity of information or by hours' work required — than a credit application, part order or insurance policy.

Many of the re-engineered processes in this briefing might be arranged easily along a scale: a well-standardised process handling great quantities of regular units right at one end, and a process handling just one unit at the other. Such a scale could be infinitely graduated, but in practice it is striking that there are certain midpoint processes (such as AT&T's OLS products or Rheden's building-permits) with a character quite distinct from both the Ford and the Kodak processes.

Also, some processes have no easily definable unit at all and therefore can't be put on a scale: IBM's Latin America system is a facility for providing all kinds of information. Making its unit the 'information request' doesn't really work: the miscellany of possible information requests is too great to be well described on a chart of operations making up one coherent process; within one session of accessing information it is difficult to say where one request ends and a new one begins; and enabling fast throughput of requests is much less of an aim than giving access to richer information.

It proves convenient to distinguish four kinds of process:

● **Transaction-based process:** unit size and variety small, considering the high volume; eg Ford, IBM Credit and the great majority of well-known examples.

● **Matter-based process:** unit markedly more substantial and varied than with transaction-based, but higher volume than with project-based; eg Rheden building-permits or AT&T's OLS products.

● **Project-based process:** one massive unit; eg Kodak new camera development.

Kinds of Process for Re-engineering

Organisation, process re-engineered	*kind of process*
(from Hammer, or Hammer and Champy)	
Bell Atlantic, new customer connections	*transaction-based*
Capital Holding, direct marketer of insurance (many processes)	*transaction-based*
Ford Motor, procurement	*transaction-based*
Hallmark, new cards design, and management decision-making	*matter-based* *facility-based*
IBM Credit, credit authorisation	*transaction-based*
'Imperial Insurance', accident claims	*transaction-based*
Kodak, new camera development	*project-based*
Mutual Benefit, new insurance policy	*transaction-based*
(other interesting examples)	
AT&T, development of made-to-order electrical products	*matter-based*
Bendix, new brakes product development	*project-based*
Bow Valley, management decision-making	*facility-based*
'Consultancy firm', skills database	*facility-based*
IBM, Latin America information	*facility-based*
IBM, marketing information	*facility-based*
McDonnell Douglas, new aircraft design and documentation	*project-based*
Ontario PM, correspondence with public	*transaction-based* *matter-based*
Rheden, building-permit requests	*matter-based*
Telstra, customer complaints	*transaction-based*
Texas Instruments, capital budget requests	*matter-based*
Xerox, making strategic plans	*project-based*

● **Facility-based process:** not definable in terms of processed units; eg IBM's Latin America information system.

This nomenclature makes it possible to use *case* as the broad term for the unit of *any* kind of process in contexts where it is irrelevant which particular kind of process is concerned.

The first table suggests how to classify all the main examples — except Taco Bell — from Hammer's original article and the Hammer and Champy book, together with all those provionally labelled earlier as product-development, managerial or otherwise of interest.

The second table gives a further selection of representative examples of the four kinds of process.

Predictability and Standardisation

The four-way distinction rests in part on the observation that in some processes the units are more *standardised* than in others. There are two slightly different themes bound together here:

● **Standard process-flow.** For any process whatsoever, you can draw a diagram, with boxes for operations and workflow lines between them, and say truly that it is a description of the standard process. The question is how detailed a diagram can be made and still remain a clear, accurate, standard description.

A transaction-based process can normally be decomposed into a great many operations, each quite precise (eg 'check postcode corresponds to address'). Although there will be numerous two- or three-way decision-points, and some will feed back to early operations, the range of possible flows through the process is not so great that a diagram will take on a spaghetti or lattice form (eg with 37 boxes, each with arrows going to, on average, 25 of the other boxes). Moving across the scale from transaction-based through matter-based and product-based to facility-based, the difficulty of defining a standard process-flow becomes ever more acute: you can sacrifice detail and precision, so that the description is accurate and intelligible but brief and bland; or make it more meaningful, by recording interactions between operations

Other Representative Processes

Transaction-based processes

Government agency: handling monthly payments of unemployment benefit (allowing for any temporary work by claimant)

Magazine publisher: taking and scheduling adverts

Town: generating schedules and processing reports for sewer inspections

Bank: authorising (or not) credit card transactions

Matter-based processes

Pharmaceutical company: handling dismissals of individual employees (in a country with strong employment protection)

Science publisher: handling the publication of articles in a journal (including feedback from expert referees)

Diesel engine manufacturer: working out choice of product options to meet customers' requirements, and producing tender documents

Bailiff: pursuing bad debts through all possible stages

Project-based processes

Advertising agency: planning and designing a campaign

Mail-order company: producing a sales catalogue

Author: producing a reference book, eg a travel guide or sports almanac

Mining company: negotiating a joint-venture agreement with a government

Facility-based processes

Fast-food chain: using a geographic information system to analyse sales, decide where to locate new outlets, test-market new products etc

Police: accessing photos of criminals and information about characteristic methods

University library: accessing catalogue of books

Local government authority: storing minutes of meetings of all committees, integrating texts with statistics and maps that become tangled and confusing; or make it less objective by omitting some possible paths that are judged less significant.

● **Standard process-content**. The processing of a delivery to Ford is standard, in the sense that, if ten deliveries, identical in all essentials, passed through the process, even if handled by different people, the outcome of each would be the same. A matter-based process, as at Rheden, is less predictable. The bureaucrats try to be consistent rather than capricious, but there are ill-defined factors: some criteria entail aesthetic judgements; there may be influences from outside the system, such as protests in local newspapers; the fate of one permit request may be affected by permission given on another. With Kodak, a project-based process, there is no predictability at all, since the whole notion of developing ten products, identical in all essentials, is too awkward to conceive. The same applies to facility-based processes.

▼ The four-kind breakdown is not meant to be the last word on the subject. A more detailed analysis could avoid the assumption that certain variables tend to overlap, eg high-volume generally means well-standardised. This refinement, though an obvious theoretical avenue, would be too elaborate for nimble, practical use. As usual with management concepts, there is a balance to be struck between analytical rigour and convenient application.

It can be objected that the labels given to the four kinds of process are not perfectly descriptive in all circumstances; eg an insurance claim is perhaps not exactly a transaction. It often happens in management science that, if you make a breakdown into categories, there is no perfectly descriptive label for each category available. The best to be expected is to find labels that give a reasonable impression and are not seriously misleading. ▲

Tricky-to-classify Processes

Is it entirely true that every process that may be re-engineered is of one and only one out of four kinds: transaction-based, matter-based, project-based and facility-based? No, the claim is only that *most* (not all) processes are pretty clearly of one of the four kinds. It is instructive to examine some of the more troublesome examples:

- **Telstra's** complaints-handling process is transaction-based: complaints about an engineer not turning up or about a wrong billing are not more complex than loan applications at IBM Credit. But what about the minority of serious cases leading to litigation? Suppose a supermarket complains and sues because its stock of frozen food is ruined in a power cut: that is surely a matter. In practice, Telstra probably needs two quite distinct processes: transaction-based complaints and matter-based litigation.

- The handling of most household or motor **insurance claims** is a typical transaction-based process. But a claim on a zoo's liability policy when an elephant injures a customer is sufficiently non-standard to be regarded as a matter. And sorting out claims arising from one aviation disaster will be so complicated that it becomes a project-based process. For an insurance company that underwrites all these kinds of risk, it is probably a mistake to think that there is one claims process that can be re-engineered; a more fruitful approach is to recognise several claims processes with distinct characteristics.

- Suppose the matter-based process at **Rheden** were taken unaltered, but scaled up by a factor of 50, for some large city with far more building-permit requests. Much greater volume implies more standardisation. Would the process then be matter-based or transaction-based? The best answer here is that the scaling-up premise is implausible. In a large city it would probably be better to have three distinct processes: transaction-based for simple cases, eg turning a small restaurant into a shop; matter-based for moderate cases, eg knocking down a cinema, and building a row of five houses instead; and project-based for huge new developments, eg converting a redundant railway station into a large shopping mall with skyscraper offices above.

- The correspondence process of the **Ontario PM** is perhaps the trickiest of all. With 30 people handling 300,000 items a year, it is safe to assume that this is largely a transaction-based process.

Kinds of Process: Decision-making Logic

There are four main
kinds of process that
may be re-engineered.

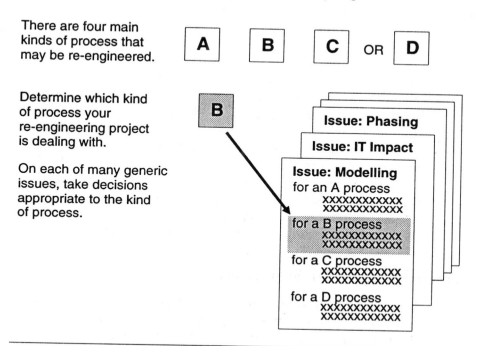

Determine which kind
of process your
re-engineering project
is dealing with.

On each of many generic
issues, take decisions
appropriate to the kind
of process.

But the occasional case dealing with specific circumstances may
have all the characteristics of a matter: eg a protest against an
injustice arising from some anomaly in the social-security regula-
tions. However, with this process, unlike those in the points
above, it seems difficult to make a clearcut distinction between
two kinds of case. The article on this case-study doesn't go into
this point.

The above examples are chosen for their trickiness. Even with
them, the four-kind breakdown brings some valuable insight.
Why, in general, should this breakdown be considered useful? As
the diagram suggests, knowing which of the four kinds of process
you are dealing with should help you make better decisions about
certain generic issues. Establishing that a process is matter-
based may help decisions on how much detailed process modelling

is appropriate; and make it easier to see certain options of implementation phasing; and so on. That is how the breakdown of the four kinds of process is used in the other briefings of this book.

▼ This appeal to decisions is an important test of any proposed classification in IT or business management. The paper money of generic categories (categories of computer applications, marketing strategies, management styles etc) should ultimately be cashable in the coin of practical decisions. Otherwise, why should anybody value it? A helpful generic breakdown will support reasoning along the lines: 'Here we have a typical type-a situation, and we should take decisions accordingly; were it a type-b situation, some other decisions would be appropriate...' Conversely, categories that, though sound and valid, have no utility in helping anybody decide what to do are sterile and worthless. ▲

Categorising by Business Effects

The other major theme among the representative ideas is the breakdown of re-engineering projects by their business effects. Here is a merged summary of the material given earlier. First, any re-engineering endeavour can be located in one of these three categories:

● the new process is worth having, but it has no great effect on market positioning;
● the new process is essential to keeping up with the leading pack in the market;
● the new process is a bold move to gain market dominance.

Second and separate, there are these three categories too:

● re-engineering to do the things currently done much better;
● re-engineering to do things not currently done, though already done by others, ie to break into new markets, as in Hammer's Progressive example;
● re-engineering to do things arising from a new definition of what needs to be done; eg the 'materials management' service outlined by Short and Venkatraman.

Any combination is conceivable, though some are more plausible than others. A process could even be in the 'worth having' category of the first breakdown, but in the 'new definition' category of the second — if (say) it was small-scale in some remote corner of the market.

How can such categorisation promote better decisions? It seems vulnerable to the line of criticism already directed against Edwards and Peppard. Choosing between the categories given might help clarify the broadest aims of a project, but it is hard to see what more specific decisions, under what circumstances, would be influenced. The analysis seems too broad-brush to give insight into the generic issues discussed in the briefings that follow. By contrast, the distinction between the four kinds of process does turn out to be consistently useful in understanding options for decision.

▼ This briefing has tried carefully to evade one source of confusion: Is a re-engineering project defined by its aims or by its achievements? Suppose a project makes drastic changes in a process with the *aim* of achieving great benefits, but the benefits don't come; should that count as a re-engineering project, albeit unsuccessful? The confusion between hoped-for results and actual achievements can lead to shoddy definition-making and muddled reasoning, not only here but in debates about the use of IT to gain competitive advantage, or about change management through empowering employees, and on many other topics too.

The main point of defining a class of re-engineering projects and finding breakdowns within it is to take a first step towards good decisions about *new* projects. When a project is new, its aims can be defined, but its achievements are as yet unknown. Therefore if, despite the above, you do want to use a breakdown by business effect, it is better to classify projects by their aims than by their achievements. ▲

CONNECTIONS

It is often implied that a re-engineering project, by definition, should address the whole of one, but only one, of the processes of an organisation. This raises the large question of how to take decisions setting the *scope* for a re-engineering project — the subject of Briefing 3.

Some of the best-known examples of re-engineering depend on the use of IT, and, to read some authors, it may appear that innovative use of IT is a defining trait of re-engineering. However, many people dispute that hotly. The controversy over the role of IT is held back till Briefing 8.

How radical must change be to count as re-engineering, as opposed to mere improvement? Briefing 11 argues that this question is about rhetoric as much as reason.

The four-way breakdown of kinds of process is used in most of the briefings. On many issues for decision it turns out that for a transaction-based process certain options and tradeoffs typically arise, while for a matter-based process the factors are usually rather different, and so on.

3. Project Scope Decisions

ISSUES

Most people who have really thought about it agree that to re-engineer all the operations of an organisation in one huge change initiative is rarely wise. That being so, the issue of scope arises: Which particular chunk of the organisation should be picked for radical re-engineering?

Process and Scope

On this issue the best start is to sketch out a forthright, straightforward procedure:

- The operations of an organisation can normally be analysed into fairly distinct chunks, called processes. An outline map of these processes need not be a source of great controversy, even though it is very different from the hierarchy shown on the organisation chart. Shrewd analysts working for a week or two can normally draw a map that will command the assent of those who know the organisation.
- The scope of any re-engineering project should normally be the whole of one and only one process. You may want to re-engineer two or three of the processes shown on the map, but then you will set up two or three separate projects. Thus the process-map serves as a menu calling for decisions.
- You decide which of the processes on the map most need re-engineering by asking obvious questions of each. Is this process in a terrible state? Or, if greatly improved, could it bring large benefits? And so on.

● Within the chosen process, the re-engineering team has *carte blanche* to consider any possibilities and take account of any factors it desires.

● Thus the scope of a re-engineering project is the whole of exactly one process from the organisation's process-map.

The first problem, as Bertrand Russell often put it, is to see that there are problems with this straightforward procedure. Take Imperial Insurance, one of Hammer and Champy's main examples. There 'accident claims' is identified as one of the company's business processes and made the scope for a re-engineering project. What could be wrong with that?

In the brainstorming scenario of the book the idea comes up of having one person, a 'case manager', responsible for all stages in the processing of any one claim. But what about having a 'customer contact', responsible for *all* aspects of a customer's dealings with the company — not just one claim but all claims, and new policies, and amendments to policies too? That idea is not debated in the brainstorming chamber, and if it were, could not be examined properly by the re-engineering team, because the repercussions go far beyond their defined scope: the claims process. Perhaps another team is re-engineering the policy process, but its scope is also too narrow to evaluate the 'customer contact' idea properly. Thus the scope decided for this company's re-engineering projects makes it impossible to discuss and develop at least one of the large, plausible innovations available.

The snag with the straightforward model outlined above is that its process-map introduces plausible assumptions about the chunks of business worth re-engineering, without recognising that they *are* assumptions and that other credible scope options may exist too. Of course, a re-engineering project must have *some* defined, constraining scope, otherwise it can't focus on anything properly — and any scope will rule out *some* possible innovation or other. The challenge then is to set a scope that is coherent and manageable, without ruling out the most promising innovations.

▼ The above line of thought illustrates a general technique helpful in exploring other management subjects or other fields of knowledge: first sketch out a straightforward solution; then probe for awkward snags; then refine the problem and its possible solutions. ▲

Co-ordinating Multiple Projects

A natural reaction to the difficulties just outlined is to oppose the notion of picking one fairly self-contained process, and re-engineering it as a project separate from developments elsewhere in the organisation. Why not first design a complete blueprint of new processes for the whole organisation, and then run staggered but coherent re-engineering projects for all the processes?

The trouble is that, however carefully processes and interfaces are demarcated on a blueprint, detailed design work on one process may still affect work on another. The co-ordination of developing designs can become a nightmare.

Suppose the claims re-engineering team at Imperial Insurance has a brilliant new idea for setting a claim's reserve (ie estimated total payout) more accurately. This idea, however, is not consistent with the overall blueprint. Is it to be rejected for that reason? Surely not.

Then some detail in the blueprint has to be changed accordingly. The amended blueprint is then passed to the teams working on all the other processes. The team for the statistics process (comparing premiums against claims) then notices that the change to the blueprint has unacceptable repercussions for its own work. An appeal is lodged. But before this can be heard, the claims team decides after more brainstorming and prototyping that the new idea won't quite work — and anyway, an even better and more radical idea has just come up, that will set up other repercussions . . .

The essential problem is this: it may be feasible to redesign process A quite radically, if the surrounding processes B, C and D are taken to be stable; but if B, C and D are also in a state of re-engineering flux, then the difficulty of ensuring that all fit together neatly may be next to intractable.

REPRESENTATIVE IDEAS

The straightforward approach is first to map out the organisation's processes, and, after that, to choose exactly one of them as the scope of a re-engineering project. Debating this approach soon raises the question of what nature of thing a process is. Unless that is clarified, a map of processes won't be a reliable support for decisions about scope.

Process: Definition and Criteria

Davenport and Short define a business process as 'a set of logically related tasks performed to achieve a defined business outcome'. They immediately give five common examples of processes. Two of them — ordering goods from a supplier, and processing an insurance claim — seem unremarkable transaction-based processes. The other three given are matter-based or project-based processes: developing a new product; creating a marketing plan; and writing a proposal for a government contract.

But what is a 'defined business outcome'? Something received by a customer. What is meant by customer? Well, not merely an external customer of the organisation, but also anybody internal who receives anything from (er) a process. Since this definition is as good as circular, it is of little use as a basis for scope decisions.

Definitions of process given by other authors and consultants are no better. For Morris and Brandon 'a (business) process is . . an activity carried out as a series of steps, which produces a specific result or a related group of results.' Question: what would be an example of a resultless activity? The more such abstractions are examined, the less helpful they seem.

Another way of capturing the notion of process is to set criteria for discovering the processes of a certain organisation, or for testing whether the process-map drawn by somebody else is good or bad. Hardly anybody offers criteria clear enough to be used. Still, there is often an unstated implication that the principle of

Occam's Razor applies: other things being equal, choose a neat map of processes over a messy one. If your draft map shows Process A connected up with Process B by numerous tentacles, alter the definition of both, to give a new map needing fewer tentacles; continue in that way until the processes are as self-contained as possible, and no neater map can be made without doing violence to the facts.

Process Granularity

Hammer and Champy favour the straightforward solution described at the beginning. (At least, they give that impression; though, if read very carefully, their text is not all that definite.) On this view, one re-engineering project should be tackle exactly one process from a process-map. Its scope should not be the the whole organisation. Neither should it be only one portion of a process. Neither should it take a couple of processes together, and treat them as one field for attention. Given all this, re-engineering can never alter the process-map of the organisation, only what goes on within each process.

The first question is the *granularity* of the breakdown. Hammer and Champy favour (or seem to favour) a map of six to ten processes in a typical organisation, as the menu from which to choose the scope of a re-engineering project. But suppose the mapping were done in a way that gave a menu of (say) 100 processes in a typical business, or that showed how some processes were made up of other processes. Then the principle that one re-engineering project should tackle exactly one process would lead to quite different scope decisions.

Andrews and Stalick say that the average business has 10 to 15 'primary business processes', and only an exceptional re-engineering project will cover more than one of them. Also, each primary process is said to be divisible into three to five 'key processes'. Unfortunately, these general statements seem impossible to reconcile with the example they give of the operation of a corporate training centre.

▼ Multiple products pose a difficulty for any abstract definition of process, or statement about how many processes there will usually be.

Cross et al give a definition of 'core process', leading naturally to the example processes of new product introduction, order fulfilment and customer service. But suppose a business has two main product lines: sausages and ice cream. Does it have the three core processes just given, or are there six? If the latter, what about a business whose product lines are rather more similar: domestic air-conditioners and military heat-exchange equipment (eg for tanks in deserts)? Or much more similar: electric kettles and electric shower units? Suppose there are seven distinct product lines; what then?

Multiple business units are also awkward. A statement that most businesses have (say) between six and ten processes raises the questions: Six and ten within a whole multi-division corporation, or within one business unit? If the latter, how do you define business unit?

Multiple products and multiple business units can muddle debate on other management theories too: critical success factors, value chain analysis or core competence, for example. ▲

Processes and Sub-processes

To pursue the concept of processes within processes, here are some things from Davenport's book:

● 'Our definition of process can be applied to both large and small processes — to the entire set of activities that serves customers, or only to answering a letter of complaint.'

● Most companies, Davenport says, have fewer than 20 major processes (major process and key process are used as synonyms): IBM has 18, Ameritech 15, British Telecom 15, Xerox 14, Dow Chemical 9. There is also a list of 12 typical processes for a manufacturing firm: product development, order management, and so on.

● Whether you should recognise three or 100+ processes in a certain organisation depends on your objective, Davenport says.

Processes and Sub-processes

*What does it mean to say that an organisation has
both main processes and sub-processes?*

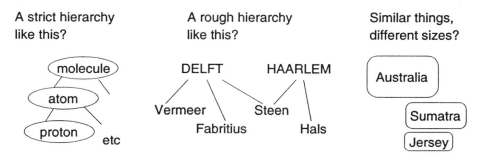

A strict hierarchy
like this?

A rough hierarchy
like this?

Similar things,
different sizes?

If it is incremental improvement, then numerous narrow processes should be mapped out, if radical process change, then far
fewer, much broader processes.

As the diagram suggests, there are at least three possible ways
these statements (and similar ones in other books) may be understood:

● *Either* different-sized processes are like molecules, atoms and
sub-atomic particles: the larger ones *are made up* of the smaller
ones in a **clear hierarchy**.

● *Or* different-sized processes are arranged only in a **rough
hierarchy** — as are Dutch towns and painters. Most 17th-cen-
tury painters worked in just one town, and for each main town
the native painters can be listed, but a minority of painters who
did substantial work in several towns spoil this neat arrangement.

● *Or* different-sized processes are like continents and large islands and small islands; ie they are separate things of the same
kind, but of **differing size**.

Davenport says that Xerox has recognised 'intellectual proper-
ty management' as a process, but not as one of the 14 major
processes. Is it then a sub-process *within* one of the 14? Or does

it *overlap* several of the 14? Or is it *separate* from the 14? Like some others, Davenport is not crystal-clear about this elementary point. However, at one moment elsewhere he seems to hold the second (Dutch towns and painters) position because he states that one relatively narrow process may cut across two broader processes.

'Key' or 'Core' Processes

Johansson et al are afflicted by other kinds of contradictions — worth untangling, since they have wider relevance for decisions of scope.

Their argument starts out like this. A business has many processes; divide them up between core processes and non-core processes. Core processes are more important to the success of the business than the others. There are typically five to eight core processes. For an insurance company, the rate-setting process (ie determining premiums by studying past claims) is an example of a core process. Re-engineering may tackle either kind of process, but the rewards, and also risks, are likely to be greater with core processes.

A natural rejoinder is that rate-setting is certainly important, but are not almost all processes (apart from perhaps the staff canteen) in an insurance company important too? Agents' accounts, statistics for industry regulators, life-fund valuation, negotiating stop-loss re-insurance treaties, investment of premiums, and a dozen others? But if all such candidates are counted, there will be many more than eight.

One possible interpretation is that any of the things just mentioned *could be* core processes in any insurance company, but the processes that actually are core for any given company are changing all the time. Thus, in one company now, rate-setting is a core process, since it is in poor shape, and improvement is essential. In two years time, newly restored to health, rate-setting will no longer be a core process. But by then other insurance companies may have developed sophisticated telecoms links with their agents, and thus, for the first time in decades, agents' accounts may be a core process. This line of speculation illustrates the confusion about the concept of the *core process*:

● One meaning is that a certain chunk of the company's activity is **inherently** a core process, by the very nature of the business. Johansson et al imply this by saying without qualification that rate-setting *is* a core process of an insurance company.

● The other (contradictory) notion is that a certain chunk of the company's activity is **at a certain moment** particularly sensitive, and thus a core process. Johansson et al also imply this by saying that the views of customers should influence recognition of core processes. If customers want shorter cycle-times, then order entry may well be a core process; if not, not. What Johansson et al don't say is that customer opinions change over time; if cycle-time is slashed dramatically, customers may start asking for improvement in some different area; thus, on this logic, order entry will fade out of the list of core processes.

▼ Had Johansson et al given some *negative examples*, had they given and explained examples of (say) five core and also five non-core processes in an insurance company, their thought could have been much more lucid. Negative examples are valuable not only in management science, but in practically any field where definitions are needed. When defining X, give examples not only of typical instances of X, but also, if necessary with explanation, of things that are not-X. After defining (say) 'cathedral', mention some large churches that are not cathedrals and say why. ▲

Do Processes Really Exist?

Hammer and Champy talk stridently about focusing on processes rather than tasks or functions, as if processes were, so to speak, real things, there to be discovered. If they are, then just as a dozen competent chemists given a certain molecule should agree which elements are in it, so a dozen competent consultants examining any organisation should all draw, at least broadly, the same map of its processes.

They don't put it quite that strongly but the assumption is implicit in much they say. Take a statement like: 'The only absolutely essential element in every re-engineering project is that it be directed at a process rather than a function.' This is practically meaningless, unless a process is regarded as a thing more or less objectively verifiable. Again, they say that once the map of the processes is drawn people should exclaim 'Of course, that's just a model of what we do around here.' There is certainly no indication that two shrewd consultants might come up with significantly different process-maps, that both could have good arguments, and that choice between them might require very fine judgement.

Many others also talk as if the processes are really there to be found. This may sound an abstruse philosophical point, but it is of the first importance to scope decisions. To see why, first consider a few exceptional statements which imply that the process is a far less definite thing:

● At just one point, Davenport and Short seem to take a line inconsistent with the rest of their article. A skilled process consultant, they say, might treat everything concerned with meeting customer orders as one process for redesign, but then again, the consultant might decide to break it into three processes — negotiating, receiving and fulfilling orders. This seems to hint that a process is not a definite thing that is there to be discovered, but something more like whatever a shrewd judge considers to be a suitable chunk of problem to tackle, given all the circumstances.

● Morris and Brandon say at one point: 'The scope of each process is important only in that it should be a convenient unit to analyze, change, and manage.' This should mean that there can be no firm rules about how to delineate processes, or generalisations about how many processes the average business has, since the unit convenient to analyse, change and manage will vary greatly from one organisation to another, from one time to another. Many of their procedures do seem consistent with this view — though the book as a whole is not very clear.

● Obeng and Crainer suggest, in contrast to Hammer and Champy, that it is often very difficult to map the main processes of an organisation. The 14-process example of their own business school backs up this judgement by being moderately unobvious.

But they don't follow through to ask the fascinating questions that now arise. Can there be several possible, equally valid ways of mapping a certain organisation? If so, how do you choose between them? Should it depend on your purpose? But can you decide your purpose before you have a map of your processes?

● Davenport's book also contains another, subtly different, view of process. Statements about process such as 'simply a structured, measured set of activities designed to produce a specified output for a particular customer or market' and 'a specific ordering of work activities across time and place' may seem innocuous, but later the notion of process is strongly identified with that of *structure*. Thus a number of loosely related activities, eg various research projects in the agricultural chemical industry, may be treated together for the first time as one process, ie given a better defined, more integrated structure. Again, much management behaviour is said to be so unstructured that the aim should be to make it more process-like, which seems to be synonymous with more structured. If taken very far this view would mean that the processes of an organisation are not necessarily already there, but may be created or reshuffled by the imposition or adjustment of structuring.

All these scraps of thought tend to undermine the notion of first making an objective map of the organisation's processes, and, after that, choosing exactly one process per re-engineering project. But they are carefully gleaned exceptions to the general run.

Choosing the Process to Re-engineer

Once you have some menu of processes, the question is which to choose as the subject of a re-engineering project. This decision is crucial, but not necessarily very demanding. Hammer and Champy give three sensible criteria, but Davenport's four are slightly better (in paraphrase):

● how **important** the process is in relation to the overall strategy of the organisation;

● how **unhealthy** the present state of the process is;

● how favourable the **change-management** factors are (eg availability of an effective, committed sponsor; positive or negative attitude to change among the people concerned etc);

● how manageable and **coherent** a re-engineering project would be (a process is, by its nature, reasonably self-contained, but still some processes have more complex interactions with the rest of the organisation than others).

Sometimes the choice of process may be obvious, but not always. If a process is unhealthy (and thus ripe for re-engineering), it may also have unfavourable change-management factors (and thus be risky). Tough decisions about awkward tradeoffs may be needed.

The article by Hall et al describes an insurance company whose approach was (arguably) somewhat different. It considered three possible *objectives* for the business as a whole: improved claims processing; more knowledgeable service representatives; and offering a broader portfolio of products. On researching customers' feelings, the company found that the first of these three was the most popular. This data was a strong influence on the decision to re-engineer the claims process rather than other processes.

Incoherent Scope-setting

Some book-length methodologies for re-engineering have such difficulty with scope-setting that they can only be called incoherent. For example, Cross et al give an overview account of a whole methodology in three main phases: analysis, design and implementation. According to this you decide within the analysis phase which processes to re-engineer in the design phase, and with what priority. But the detailed description of the analysis phase is with respect to one process only (which must have been chosen already). Then at the start of the design phase you are told to define and get agreed the scope of the project, ie the boundaries of its process; the authors don't say how you should do this, only that you should.

The book by Petrozzo and Stepper has 60 pages on scope-setting and gives seven potentially relevant factors, but it doesn't say how to identify what processes there are in the organisation, nor how to choose between rival processes for re-engineering.

Andrews and Stalick are not very clear either. The first of their standard steps seems to set the scope for one particular re-engineering project, and the second step to define a vision for the process that is covered by the project. And yet in the extended script provided of a typical workshop to develop a vision, the debate seems to be at the level of the organisation as a whole, not the specific process.

As this shows, there is often confusion between two distinct things: debate at the *organisation level* before the decision is taken to re-engineer (say) processes A and B, but not C, D, E etc; and work to be done at *project level* when process A is being re-engineered. It ought to be easy to keep the two things clear, but these three books all fail.

▼ The question looms: What can possibly be learnt from mere criticism of the the methods of others?

First, attractively produced books from reputable publishers that seem to contain all the usual re-engineering themes, often turn out, on examination, to be confused and feeble.

Second, the scope-setting issue is one of the best tests of the quality of any book or method or consultant.

Third and most fruitful, the confusion on this issue results from a conflict of factors well worth exposing. On the one hand, it is *attractive* to begin with substantial investigation at the level of the whole organisation; the more you do, the better placed you are to make well-informed scope decisions about re-engineering projects. On the other hand, doing a great deal of organisation-level analysis is *unattractive*, because (since you will probably re-engineer only a minority of the processes) much of the detailed work will be wasted. Unless the conflict between those two factors is firmly grasped, confusion may ensue. This is an example of the benefit from examining recommended methods for management and IT critically. Finding contradictions and confusions helps you clarify your own thinking about the trickiest issues. ▲

Contrary Positions on 'Process'

One common position

Draw a high-level map of the organisation's processes

Then pick (say) two of them for two re-engineering projects.

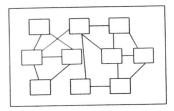

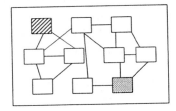

An opposing position

Draw a high-level map of the organisation's processes

Then decide appropriate scope for (say) two re-engineering projects.

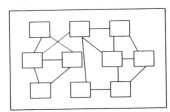

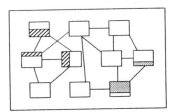

DISCUSSION

There is more at stake here than academic quibbling about how to define 'process'. As the diagram suggests impressionistically, there are two main contrary positions whose implications affect scope decisions. This section examines the arguments.

Two Positions on Process and Scope

Many writers seem to hold the following position, at any rate implicitly:

● It is possible, and desirable, to draw a process-map of any organisation based on certain clear principles that define the notion of process. That being so, most competent analysts will draw much the same process-map of any given organisation.

● Scope decisions choose the process(es) from this map most in need of re-engineering. It is best to have one project for exactly one process (no more, no less). Other possible decisions (eg re-engineer half of process A together with half of process B) don't arise for consideration.

● Thus, in the customer-orders example cited by Davenport and Short, if your initial mapping showed that 'meeting customer orders' was one process, you would either set up a project to re-engineer that whole process, or no project at all.

The contrary position, just hinted at by a few authors, is the following:

● Process is a vague term: too vague to support the method of drawing a process-map first, and after that, choosing exactly one process to re-engineer.

● The shrewd course is to look at the company's set of operations as a whole, without any assumption that it 'really' contains certain processes, and to ask: What would be a suitable chunk to re-engineer?

● In the customer-orders example you might arrive at a scope decision after considering a number of scope options: A, negotiating, receiving and fulfilling orders; *or* B, as A but also including after-sales service; *or* C, negotiating orders only; *or* D, negotiating orders together with promotion campaigns; *or* E, as A but only for industrial customers and products (not retail stores); *or* F, as A, but with two re-engineering projects, so that two main classes of products will be handled quite separately from now on. In this decision-making, any arguments based on the notion that certain operations 'really' belonged in one process or another would be pointless distractions.

▼ Test any kind of breakdown by asking: Is this meant to be *factual*, something that can be verified (eg 'the USA consists of 50 States'), or just *helpful* (eg 'there are five main dialect-areas of American English' — or nine or 31, depending how fine an analysis is needed). Both types can be useful, but you need to know which of the two you are dealing with. 'There are nine departments in this company' is a factual statement that may be true or false. But, on the argument just given, to say 'there are seven processes in this company' is to say no more than 'one convenient way of looking at this company is to divide it into seven pieces.' Somebody else may find a breakdown of the company into thirteen processes better, and so it may be — but only in the sense of more elegant or convenient or helpful. That is entirely different from holding that one breakdown is true and the other false. ▲

Comparing the Two Positions

The first position has the advantage of making decision-making much simpler, but the greater disadvantage of using flimsy premises to derive implausible conclusions.

If it is feasible to start out by drawing a map identifying processes in an objective way at the start, why has nobody set out any workable principles for doing so? Why do some of the examples in the most influential publications contradict each other? Davenport and Short give the ordering of goods from a supplier as one common example of a process. But Hammer and Champy make a large point of saying that the whole of procurement at Ford (purchase order and accounts payable and goods inward) is one process.

Again, many writers give product development as a good example of a process within a manufacturing company, but one article, though admittedly not famous, avers that the truly go-ahead company will re-engineer one whole 'product realisation' process, including marketing and servicing as well as product development. This is too sweeping to be a plausible generalisation, but neither can it be discarded as never sensible for any company under any circumstances.

Moreover, the common view that the chunk of the business chosen for re-engineering should bring together operations hitherto walled off by departmental boundaries can easily be overdone. If an insurance company has a claims process handled mainly by the claims department, then some interaction with the accounts department and perhaps the agents' liaison department is indeed necessary. But that is true of any claims process whatsoever, including a computer-based process of 25 years ago or a ledger-based process of the last century. It isn't true that, prior to re-engineering, systems never crossed departmental lines. Note also that the smaller the process to be re-engineered (eg one process out of 100), the less likely it is to span departments.

Recognising the Tradeoffs

If the first position were valid, it would follow that the scope options (A to F) suggested above for the customer-orders example were pointless and not worth considering for a moment. But if, as seems likely, at least some of these options would be worth debating, then any doctrine that prevents them even surfacing must be unsound. Surely the wise approach is to accept the second position, and go on to recognise this tradeoff:

- Radical re-engineering of the whole organisation in one project is much too ambitious. Therefore it is essential to seize on some convenient chunk. But a chunk can be hewn to any granularity. Purchase order is a convenient chunk that brings together dozens of more detailed operations. But it is itself only one part of a larger chunk called procurement, and that is itself part of a manufacturing chunk. One (not the only) challenge in defining re-engineering scope is picking a chunk of suitable size.
- Picking a relatively large chunk to re-engineer is ambitious, and thus risky, and probably incurs large investment costs. Still, the larger the chunk, the more scope there is for sweeping changes and spectacular benefits.
- Picking a relatively small chunk reduces risk and investment cost, but perhaps at the expense of some opportunities for extensive changes and benefits.

● The aim must be to find the chunk that provides the least-bad balance between these contrary factors. This calls for situation-specific judgements.

Two Examples of Scoping Dilemmas

Given all this, the idea that you should *always* choose one out of the six to ten easily identifiable main processes is not sustainable. Hammer and Champy's Imperial Insurance, for instance, has far more options than at first appears. The accident claims process is introduced as if it were one of the obvious six to ten processes in a motor insurance company. Perhaps so — on a map of that granularity, but it isn't necessarily a good scope for a re-engineering project. Shouldn't the company consider some other scope options — such as those suggested in the diagram?

To take underwriting and claims together gives a very broad scope, but could be the only way of catching perhaps the most striking innovation of all: having a 'customer contact', responsible for *all* aspects of a customer's dealings with the company — not just claims but new policies and amendments to policies. A much narrower scope might include commercial (as opposed to personal) claims only; perhaps the handling of personal claims is already adequate, and the best target for innovation is claims on fleets of vehicles insured by companies.

Remember too that Imperial Insurance is a narrow example, because only motor insurance is handled. A company with household, personal accident, liability, fire and other kinds of insurance business would have many more scope options to consider.

The correspondence process in the office of the prime minister of Ontario provides another instructive example. As described by Proc et al, there was a classic scope dilemma: re-engineer just that process, or make an integrated, government-wide process, covering correspondence in the other 26 ministries as well.

There are attractive arguments for the second option: the PM's office often needs to pass on letters it receives to a ministry for attention, and vice versa; it is desirable to avoid contradictory

Scope Options at Imperial Insurance

Plausible decision-making

High-level map of the organisation's processes

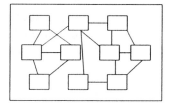

Choose the whole of this particular process as the scope for one discrete re-engineering project

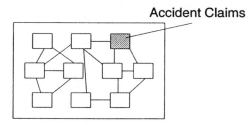

Accident Claims

But why not consider other scope options too?

Broader

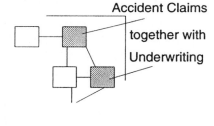

Accident Claims

together with

Underwriting

Narrower

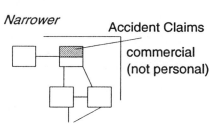

Accident Claims

commercial (not personal)

Different

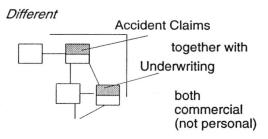

Accident Claims

together with

Underwriting

both commercial (not personal)

statements by the PM's office and a ministry; and so on. But the first option has the great advantage of being less ambitious, and therefore quicker to do and more likely to succeed. Moreover, it allows ministries to try out a variety of methods and technologies, so that over time the less successful can wither, while those that work well are extended.

Plainly, this was a crucial dilemma of scope. Nobody can judge at this distance whether sticking to the PM's office only was the right choice. This kind of decision calls for delicate judgement of the situation-specific factors.

Kinds of Process, Scope Decisions

The range of options for scope decisions tends to differ with the *kind of process* concerned.

● With a **transaction-based process,** scope decisions can be particularly exacting and significant. As Imperial Insurance illustrates, transactions may differ along at least two dimensions: what they are about, eg policy or claim; and what kind of business or market they concern, eg motor or household, commercial or personal. In many organisations there are quite a few ways of analysing transactions, and thus many plausible scope options to choose between. As with Imperial Insurance, the only way of deciding the scope may be to peer into the likely content of design discussions (eg Is the 'customer contact' role a likely issue?) — a tricky, annoyingly unmethodical, and perhaps time-consuming procedure.

● In a **matter-based process,** the unit of process is by definition a larger, more substantial thing than a transaction. This tends to reduce the number of scope options and thus make scope decisions easier. A matter is more likely than a transaction to correspond to one obvious, natural chunk of the organisation's work.

● The scope of the **project-based process** is, in general, the easiest to demarcate, since it usually consists of everything necessary to make that one defined product at which the project is directed.

● The **facility-based process** is often very tricky to scope — but in a different way from the transaction-based process. It may be easy enough to point to the process on an organisation-wide map and mark it off from the rest of the organisation's activities, but the scope uncertainty may lie in the depth of the facility's coverage or the degree of its sophistication. A paragraph of text may define (say) a consultancy's skills database, but this may allow almost infinite gradations of amplitude. Should the content of documents produced by consultants be stored? Literally all documents, including training courses and books? What about standard documents issued in slightly different versions on different occasions? How sophisticated should the search facilities be? Including keywords, hypertext links, expert-system-based extraction of information, neural-network-based mining for significant patterns? With this type of process, a sophisticated system could be more ambitious than a simple one by a factor of 100. It seems wise to agree some rough definition of scale before turning the re-engineering project team loose. But that kind of definition, to be soundly based, may call for substantial intensive, highly-skilled study first.

REPRESENTATIVE IDEAS

This section discusses the strategy of running multiple projects to re-engineer a considerable portion of the organisation's processes.

The 'Exhaustive' Strategy: Strong Views

In their influential article Davenport and Short contrast two strategies for re-engineering:
● With the **exhaustive strategy** an organisation makes an overall plan to re-engineer all its processes. Normally, it won't be sensible to do everything literally in parallel; priorities must be set. Nevertheless, all the designs developed will be held consistent with some organisation-wide blueprint.

● The **high-impact strategy** is to concentrate on re-engineering just a few carefully chosen processes.

These authors are positive that the high-impact strategy is, in general, resoundingly superior. In his book Davenport gives three powerful reasons that the exhaustive strategy is usually ill-advised: the cost is too high in funds and human effort; the psychic shock and upheaval are too great; and the co-ordination of all the change projects is too formidable. He mentions that IBM is one company trying to defy this logic, and hints that IBM may not be wise, or that it may end up with only minor rather than radical changes throughout the organisation.

Hammer and Champy take the same general view. Theoretically their straightforward process-map approach would allow simultaneous re-engineering projects for each one of the processes, but they say unequivocally that no company should ever do this. Again, interviewed for Keenan's article, Hammer says that it is rarely advisable for a company to re-engineer everything at the same time.

With this body of opinion on one side of the argument, why should the exhaustive strategy even be worth attention? As might be expected, most of the examples in Hammer and Champy's book are consistent with the authors' advice — but not all. The expansive statements quoted from the people at Hallmark and at Capital Holding, if they are anything more than bombast, can only mean that those companies intend to re-engineer most or all of their business processes in a tightly integrated way.

The Capital Holding Example

Capital Holding, a direct marketer of insurance, decided to take the company's existing sales, service and marketing processes apart, and put them back together again in a new configuration. To be more precise, it undertook the massive task of re-engineering all its processes in a coherent, co-ordinated way. After a radical new model of the company had been sketched out, a dozen teams were set up, each to work on one of the main processes of the model.

It is of the first importance that in this example, unlike almost all others in Hammer and Champy's book, the exuberant rhetoric is not accompanied by any claim that any tangible success has been achieved. There are no dramatic statistics about reduced costs or cycle-times. The description given is of plans the company has made and things it hopes to achieve over the years ahead. These aspirations may be imaginative and visionary or naive and reckless; time will tell. The real value of this case-study is that it exposes some of the difficulties with staggered re-engineering projects that together cover all or most of an organisation.

Capital Holding needs about a dozen separate projects for detailed redesign work, since otherwise the scope of a project would be too large to be manageable. How are the process designs developed on these projects to be kept consistent? By reference to an overall business model, which shows how the processes are meant to fit together.

But, the people at Capital Holding report, 'We also learned that you can't plan an entire re-engineering project in advance, because what you discover during the project changes your plan. Every change you design is a living, rough draft, not a perfected process. Re-engineering is an iterative process.'

Does this mean that the co-ordinating business model itself is in a state of flux? If so, maintaining consistency between the model and the designs produced on all 12 projects will be a hard task. Or is the business model so bland and shallow that it can survive unaltered, no matter how the process designs develop? If so, how can it play the role of co-ordinating the work done on separate projects?

This problem is inherent in any attempt to re-engineer an organisation exhaustively. But Capital Holding piles on another aspiration: 'We're building a flexible system in modular parts. If one part becomes irrelevant in a year, we can throw it away. If we can't build a system that we can keep changing over time, then we didn't do the right job up front with our vision.' This imposing signpost could be pointing the way to a thicket of difficulties.

More Concepts and Examples

An article by Talwar points out correctly that one process re-engineered as a discrete project may not necessarily fit in well with other processes, and it goes on to enthuse about the exhaustive, top-down strategy. But behind the blizzard of buzzwords (strategic focus, mission, intent, values, vision of re-engineered organisation) there is no treatment at all of the essential problems. At what point should organisation-level design work cease, and split into process-specific projects? By what arrangements should they be co-ordinated thereafter? These topics are not mentioned. But they are fundamental. Unless you take sound decisions about them you will have no credible strategy.

Cavanaugh's article gives detailed accounts of re-engineering at five American power utilities. In at least four of them, re-engineering has been taken to mean redesigning the whole company in an exhaustive, integrated way, as opposed to focusing on a few high-impact processes. This article concludes, poker-faced, with a management consultant's opinion that the best approach for any utility is to re-engineer one process at a time — in other words, that the companies described have got it all wrong.

The article by Caron et al about re-engineering at CIGNA Corporation describes enthusiastically how an 8000-person insurance company was given an entire set of new processes. After a master-plan had been made, projects to design six processes in detail were run in parallel. But this article too is silent about the things most worth knowing. Do these six processes represent most of the master-plan, half, 10%? How interconnected are they? How detailed is the master-plan? How exactly does it define the interfaces between the separate processes?

Lynch et al describe how Melbourne Water produced a new ten-process design at organisation-level as the first step in organisation-wide re-engineering, but as with Capital Holding, on a careful reading, it turns out that the company hasn't yet *implemented* much in the way of new working processes; in fact, some of the system development work already done has been halted and frozen. Thus, this case-study, at first sight so full of detail and rich in diagrams, provides no insight at all into the problem of co-ordinating multiple projects in parallel.

Ask some of the large management-consultancy companies about re-engineering, and the exploits they boast of are often sweeping changes throughout large organisations, that, as with CIGNA or Melbourne Water, must have required many interrelated re-engineering projects. But they don't explain how such ambitious changes were co-ordinated.

▼ Nor can you judge whether the projects in the consultancies' marketing literature are typical. This is an instance of the universal truth that, although consultancy companies are generally in the forefront of application of new management theories and techniques, it is virtually impossible to get hold of substantial information about what they do and how. ▲

An Organisation-wide Methodology

The book by Morris and Brandon is arguably not about re-engineering at all as most people understand the term. Its methodology seems to favour parallel, strongly co-ordinated re-engineering projects. Rather than pick one chunk of the company and change it radically, you start with an organisation-wide documentation exercise, and thereafter keep track of all ripple changes that may affect anything anywhere in the company. The standard methodology is as follows:

● Positioning — documenting the current systems of the **whole organisation** and associated data (eg market strategy). This entails drawing workflow diagrams of all the systems; the granularity is undefined and no examples are provided, but it does seem quite a large task.

● Re-engineering (steps 1-3) — work at **organisation level** to decide on the scope of one or more re-engineering projects; step 1 deliverable: selected possible projects worth considering; step 2 deliverable: selected projects with impact analysis on existing systems etc; step 3 deliverable: list of definitely chosen projects, with defined scope.

● Re-engineering (steps 4-9) — applied to each **re-engineering project** separately: design the new system and implement it. How the design work is kept consistent with the positioning documentation is not really explained.

Will this work? Who can tell? Virtually no examples are given.

DISCUSSION

Ambitions to re-engineer the entire organisation are understandable but usually naive. This section explains the key problems that are so tricky.

Comparing the 'Exhaustive' and 'High Impact' Strategies

Davenport and Hammer and their colleagues favour the high-impact over the exhaustive strategy, and most experienced management consultants will probably share their judgement. It is easy to see that, of the three considerations mentioned by Davenport, measurable costs and psychic-shock, will often, though perhaps not always, be decisive in themselves. However, the third factor, *co-ordination*, deserves fuller exposition.

If one chunk (A) of the organisation is radically redesigned, the resulting new design may not fit back neatly into the machinery of the organisation as a whole. Why? Because A won't be absolutely independent; there are bound to be some interactions with chunks B, C, D etc. The answer is, first, to define chunk A as cleanly as possible, minimising interactions with other chunks; and second, to reduce complications by keeping B, C, D etc stable, ie not subject to drastic re-engineering too. This may not be an ideal solution, but it is normally the least-bad. Consider the alternative, the exhaustive strategy.

Work on radical new designs for all chunks A to J of the organisation in complete isolation, and the job of fitting them all together afterwards will probably be hopeless. Conversely, treating the whole organisation as one huge chunk, subjected to one

Exhaustive Re-engineering: the Difficulties

Concept:
Design an organisation-wide blueprint,
relating all the new process designs together.

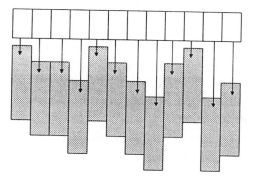

Blueprint done first.

Separate processes
developed and
fitted together in
semi-parallel.

Problems:

1. It is extremely difficult to make such a blueprint that is innovative and coherent and elegant.

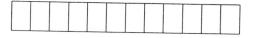

2. In practice, bright new ideas will arise within one process, with implications for other processes.
Change control procedures become complex.

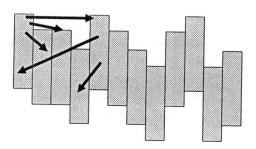

detailed redesign, probably won't work either, because the project will be too large.

Therefore, as the diagram suggests impressionistically, the only remotely credible way to pursue the exhaustive strategy is to accomplish two very demanding things. You need to begin with a large redesign effort at organisation-level — not so detailed that it is unmanageable, but substantial enough to provide a robust organisation-wide blueprint. How substantial is that? Enough to define the boundaries and interfaces between the chunks that will be tackled as separate projects.

Second, you need to devise some organisational mechanisms (project liaison representatives, joint project committees, staff-function re-engineering control department, change control forms and meetings, etc) to ensure that, as the design of each chunk proceeds into ever-more detail, it remains consistent with the blueprint — and that the (virtually inevitable) changes to the blueprint itself are properly reviewed, amplified, clarified, authorised and implemented.

This kind of exhaustive top-down approach was often tried for information (as opposed to process) engineering by organisations during the eighties. Many succumbed to paralysis by analysis, and even those that succeeded often only developed databases and IT systems to support *existing* business processes better. Development of *innovative* process designs must be much harder to co-ordinate.

In brief, there are two main issues to be resolved in considering any strategy of re-engineering several related processes in a staggered or parallel pattern:

● What kind of a thing is the initial co-ordinating **blueprint**: five pages or 105? What kind of scope does it set for the separate re-engineering projects? In how much detail does it define their boundaries and interfaces?

● What kind of co-ordination and **change-control procedures** are used to keep the re-engineering projects in step?

Unless sound decisions are made on these two issues an organisation's planning has little chance of success. Even then, the weight of authoritative opinion and argument is against the exhaustive strategy.

CONNECTIONS

A scope decision is taken *at organisation-level*. Once a certain chunk of the business has been selected as the scope for a re-engineering project, there are then decisions *at project-level* of design approach (Briefings 4-8), of process design (Briefing 9) and of implementation approach (Briefing 10).

Surprisingly many authors muddle up the issues and decisions of these two levels. Often this is just carelessness, but a couple of frontier areas do have their tricky aspects.

Briefing 5 discusses how design work on a particular project may (or may not) be influenced to some degree by organisation-wide strategic factors.

Briefing 9 is about the generation of design options for decision; the options worth attention may be affected to some degree by the organisation-level considerations that led to the scope decision to set up the project in the first place.

4. Untangling Design Approach Decisions

ISSUES

How should you go about the task of redesigning a certain process? Some clear decisions are needed. You might, for example, make a detailed model of the current process, because that seems a natural thing to do or is prescribed in some standard methodology. But it is far better to take a rational decision on the approach to process modelling that fits the circumstances of the particular re-engineering project.

How much to document the current process is one among many generic *design approach* issues, that are best addressed explicitly and call for situation-specific decisions. There are quite a number of others. In designing the process, what modelling conventions should you use (there are many possibilities and they vary widely)? Should you follow a certain multi-step methodology for re-engineering (those advocated differ, sometimes superficially, sometimes profoundly)? Should you work with a prototype version of the process? What kind of automated facilities should you provide to support the design work? How much benchmarking of other companies' processes should you do? How should you keep the process design consistent with the organisation's general strategy? How intensively should you assess the quantitative implications of the new design? Should you design in non-technical terms first and fill in the technology implications afterwards?

Taken separately none of these questions of design approach are too hard, but taken as a tangle they may become almost impenetrable. The connections and overlaps between issues are not all self-evident. The first step towards a consistent set of decisions is to distinguish the generic issues and options, and see how they relate to each other.

It is a curious thing that most advice in books, articles and brochures about how to do re-engineering is on one of two possible levels:

- *either* at the level of a **whole multi-step approach** to the design work: recommending use of a certain standard set of linked procedures, irrespective of type of process or other situation-specific considerations;
- *or* at the level of one particular **type of activity** (eg modelling, benchmarking etc): recommending that the activity be carried out in a certain way (though occasionally with a disclaimer about the exact approach depending on circumstances).

Neither of these types of advice addresses the problem of arriving at a set of approach decisions, that is both coherent and appropriate for the particular situation. The main purpose of this briefing is to untangle the issues of design approach and show the connections between them. Then the subsequent four briefings can safely probe specific issues in more detail.

REPRESENTATIVE IDEAS

The structuring of activities into a coherent re-engineering project is a complicated subject. The best way into it is to review a range of typical recommendations — not because they can be accepted as they stand, but as a stimulus to recognising the issues and options, and seeing where the difficulties arise.

Representative Multi-step Approaches

Various writers and consultancies have proposed standard steps for the work of *re-engineering design*. Five are summarised here.

In Davenport and Short's article the first two steps recommended set the scope of the project; here are the other three, concerned with design approach:

- Study the existing process, both describing and measuring it.

- Identify IT levers that stimulate new possibilities.
- Design the process, build a prototype, try it out, and refine it, until a design sound enough for pilot implementation results.

The five-step approach in Davenport's book is considerably different. The first step produces the scope decision about the process to be re-engineered; the other four are concerned with design approach:

- Identify change levers, eg new possibilities in IT and in human resources management.
- Develop process visions, ie broad-brush outlines of the new process.
- Understand existing processes, in so far as germane to the design work.
- Design and prototype the new process.

Of Morris and Brandon's whole re-engineering methodology four steps are concerned with design:

- Make a detailed model, descriptive and quantitative, of the present process.
- Make one or more possible new designs.
- Evaluate the costs and benefits of the possible new designs, and make a provisional choice of which is preferable.
- Firm up (or presumably change) the choice.

In the book by Cross et al it is unclear where study of existing processes — done to decide what process to re-engineer (ie to take a scope decision) — ceases, and design work on a chosen process starts. This is a pretty serious point to be confused about. The following is offered as a least-bad interpretation of their proposed successive steps:

- Make a detailed baseline model of the existing process, both describing and measuring it. The granularity is suggested by the example of 284 activities at a single-product insurance company.
- Produce 'design specifications', an initial rough blueprint for the new design. Judging by the example given, this is typically a couple of pages in bullet form, defining a dozen objectives, as well as various assumptions, constraints etc.
- Produce a 'high-level design' — but there is no example to illustrate its granularity.

● Produce a detailed new design — of the same granularity as the baseline one, though, since the new one is probably improved by streamlining, there may be many fewer operation-boxes (eg 284 down to 85).

● Support the new design with quantitative evidence relative to the current process; eg statistics on cycle-time, on the proportion of activities assessed as 'value-adding' etc.

● Expand the detail of the new process with 'procedure sheets' spelling out what happens in the operations shown on the model.

● Carry out detailed quantitative analysis, perhaps including simulation, to justify the new process.

Andrews and Stalick have a standard eight-step approach, but the work of making a new process design is concentrated in steps two, three and four, and then mainly in step three. Step three delivers ten main documents, ranging from a process model through to a statement of 'belief systems'. These deliverables are supposed to be produced in a series of 'facilitated workshops', each including from 10 to 30 participants, over one to two months.

▼ A methodology for any kind of design or planning task usually needs to be described on at least two planes: the deliverable *documents* it produces, and the *procedures* for making them. Many descriptions of methodologies are flawed by concentration on one of these at the expense of the other, or by muddling them in a confusing mixture. Moreover, it is often essential for comprehension to give representative *examples* of the documents produced. Andrews and Stalick provide tremendous detail about workshops (ie *procedures*) to develop a 'vision, values and goals statement' (the main product of their step 2), but no example of a representative *document* that might result. Without this, it is impossible to judge how credible the proposed approach is.

Andrews and Stalick make the claim that the methodology they describe is the basis for *all* successful re-engineering projects. This is too absurd to be interesting, but it does provoke a question worth asking of any multi-step methodology — whether in re-engineering or any other field: What reason is there to believe that this methodology is superior to a dozen others that are available? ▲

Iteration

These approaches differ in obvious ways: Davenport and Short give more prominence than most to IT; Morris and Brandon are more keen than others on comparing options; Cross et al give special attention to quantification; and so on. But all five are unsatisfactory on one vital subject: the *iteration* between the activities they recommend.

An approach in (say) four successive steps, each completed definitively before the next starts, is quite another thing from one in four steps, where the intention is to loop backwards and forwards between the steps, revising designs already made, scrapping ideas that seemed attractive but on further investigation don't work out, and so on. In fact, the choice between a strongly *iterative* approach, and a strongly *methodical* one, where a ratchet closes after each step, may affect the character of the project more than the question of the number of steps there are.

Davenport and Short don't mention any iteration between their three activities; they don't suggest (say) going back from the middle of the prototype work to study existing processes again.

In Davenport's book it is obscure how firmly sequential the four recommended activities are intended to be ('Although the sequence of the activities may vary, aspects of the ordering are important.'), still less whether iteration is to be encouraged (eg going back from the understanding of existing processes to identify more change levers).

Since nothing is said on the subject, the presumption with Morris and Brandon's approach must be that (except in the event of gross blunders) there is no iteration between activities.

The Andrews and Stalick approach seems to rule out much iteration, though the question is never addressed explicitly. The overview diagram contains no feedback arrows going back into step two or step three.

In the more detailed methodology of Cross et al the overview diagram contains dotted feedback lines between some but not all of the main activities; but later on the text itself says that the diagram is misleading! There is talk of validating and justifying a design repeatedly, always with the stress on measurement, with the implication of feedback if necessary. This, if taken literally, limits the force of any feedback, since it concentrates on justification relative to the old process rather than on finding the optimum new process. A new design might be twice as good as the old one, but if you concentrated only on proving that, you might never discover some more imaginative variant three times as good.

Distinguishing Prototype and Pilot

The general question of iteration is brought into focus by the specific question: What does *prototype* mean? Davenport's book makes the distinction between the prototype and the pilot very clear:

● The purpose of **prototyping** is to refine a new design by trying out various possibilities. It is quite acceptable to go back and forth altering features of the design. Anyone who thinks that this is a sign of failure has misunderstood the concept of prototyping.

● The purpose of a **pilot** is to achieve success, rather than to test the quality and correctness of a new process. Anyone who thinks it satisfactory for a pilot to expose significant flaws in the design has misunderstood the concept of a pilot.

The value of clarity on this question is shown by the difficulties Cross et al run into. They talk of possible feedback from the 'pilot' to 'detailed design', and of 'freedom to experiment, tune and change the design if necessary', though with no example of the kind of changes they mean. In their view: 'At the heart of the reengineering pilot is a spirit of design, experiment, checking results, and making adjustments.' In the context of the whole methodology this apparent freedom to feed changes back from the *pilot* seems to be a misleading fudge — since it surely can't be acceptable for changes of any great moment to be fed back to overturn the detailed work (procedure sheets, justification by quantitative simulations etc) of earlier activities.

▼ Few people other than Davenport keep the distinction clear, and those that do don't necessarily use the two terms 'pilot' and 'prototype'. Anybody who aims to think properly about management or IT topics will hit this terminology problem time and again. ▲

Examples of Prototype and Pilot

As a thought-experiment, consider how the concepts of prototype and pilot might have been applied to the famous redesign of Ford's procurement process.

The design team has the radical idea of not accepting a delivery of goods unless it matches the purchase order stored in the computer database:

● A **prototype** is set up to try out the idea. In experimental use, it turns out perhaps that certain types of delivery can't be handled satisfactorily by the new concept. This suggests the notion of a two-way split: some deliveries to be handled in one way, others in another way. Further prototyping tries this out; perhaps then it isn't always easy to allocate a delivery to one category or the other. And so on . . .

● When the design seems mature a **pilot** version of the process is set up to handle real deliveries on a limited scale at one location. Experience with the pilot shows that, in a minority of cases, the clerk in the goods-inwards department has to call up three or four screens of data in order to decide whether to accept the delivery. The design is then tuned to ensure that at worst no more than two screens of data are ever needed.

The pilot is not intended to cause Ford to decide that the whole concept of refusing deliveries that don't match orders is misguided, and that therefore a large part of the new process ought to be changed. No sane manager would want the experimentation and refinement described above under prototyping to occur in a pilot context, when real deliveries of goods actually needed by the production-line were being handled. Of course, if blunders have been made by the design team, it is better that they be discovered during the pilot than later. But this is like saying that, if a new toothpaste product is poisonous, it is better to discover it when only a dozen customers have died in one test market than later on when the toothpaste is on sale nationwide.

The Ford example is worth pursuing. In their book Hammer and Champy say that when a delivery arrives a clerk checks to see whether it corresponds to an outstanding purchase order in the database. According to these authors only two possibilities exist: it does correspond and is accepted, or it doesn't correspond and is rejected. This may be the most famous re-engineering case-study of all, but the account given by Hammer and Champy can't be accurate. There are at least three possibilities: the delivery corresponds precisely to an outstanding order in every detail; *or* the delivery has no apparent connection to any outstanding order; *or* the delivery corresponds to a certain outstanding order but not precisely (eg the wrong quantity is delivered, or the goods are delivered early etc). The third possibility can't be wished away by an act of re-engineering: if 30 tons of steel are ordered, and only 29 arrive, Ford surely won't refuse the delivery, if that means leaving the production-line idle for the rest of the day.

Therefore the central issue in this part of the process design becomes: *How closely* must a delivery correspond to an outstanding purchase order in the database if it is to be accepted? The answer to this may be quite a complex set of rules; eg one factor may be how badly the goods are needed in the factory at the time. This is just the kind of design issue where prototyping and piloting come into their own. The design team can prototype a variety of plausible sets of rules for accepting or rejecting deliveries, before settling on that which seems likely to work best. During the pilot stage, with real deliveries, the rules ought to turn out essentially sound, but they may very well be tuned in minor ways.

Top-down Concepts

Iteration, for which prototyping is one technique, has an opposing aspect, summed up in the term *top-down*. Here Davenport's book provides food for thought.

With some methodolgies it is self-evident that certain activities must follow a certain sequence, because they depend on each other; (say) step 2's product is the starting point of step 3. But this is only weakly true of the methodology in Davenport's book. It is stated, reasonably, that the step 'understanding existing processes' can be more efficient if you know what you are looking for, after the earlier 'developing process visions' step. This is a loose rather than a firm link. It means only that you will be better placed to judge what is material about existing processes, not that the product of the previous stage can be expanded in some explicit procedure to work out what is worth investigating in existing processes.

But in some places, Davenport also seems to advocate a tighter, top-down style. The 'developing process visions' step is essentially top-down development through a series of workshops. For (say) an order management process, the main objective, decided early on, might be to reduce processing costs for customer orders by 60% over three years. From this follow decisions on the main attributes of the new process (eg credit-checking based on expert systems; automated proposal generation; increased worker empowerment; credit, shipping and scheduling functions all to be done by the same customer-facing individual). From this performance measures and objectives are firmed up. From this critical success factors; and so on.

Within the step 'designing and prototyping the new process', it is suggested, design can be done at three levels of detail: sufficiently compact to be intelligible for the process as a whole; more detailed, for each of the main parts (sub-processes); more detailed still (activities). Is the suggestion that the design should always have a hierarchy of these three levels of detail, or merely that there should be *some* hierarchy of detail, whose levels will vary according to the situation? The latter seems more plausible, but the text seems to suggest the former.

There is another ambiguity lurking here. Are you supposed to produce the level-1 design; then freeze it; then produce the level-2 design; then freeze it; then produce the level-3 design? Or should you make a draft design, roughly complete at all three levels of detail, and then start improving it, so that (say) a change at level 3 has repercussions affecting the design at levels 2 and 1, and these changes in turn affect other parts of level 3. Davenport's use of the term 'iterative' suggests the latter, but the text as a whole suggests the former. In any event, the point is slurred over, and there are no examples.

DISCUSSION

The rest of this briefing stands back from the specific recommendations and criticisms given so far. It charts out several peculiarly tricky areas, and provides a concluding overall chart of the field of design-approach issues.

Activities of Designing and Proving

As the following diagram shows, a number of possible activities can be described as designing a better process, trying it out or proving its validity.

• A project team normally **models** the new process. The issue of approach is usually not whether to draw a model, but how detailed to make it and what conventions to use. The technology of modelling software can support the drawing and storage of diagrams of a business process, just as the technology of CAD can support the design work for a new aircraft. In the same way that the plans of an aviation engineer are just plans not actual aircraft, business process diagrams are just diagrams, not pieces of working software.

• A model is a graphic description of a process (its operations, perhaps who carries them out etc). If an extra layer of quantita-

Activities of Designing and Proving

PROCESS MODEL

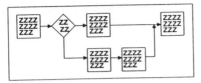

CALCULATION

Operation 1	10	1.00	10.0
Operation 2	20	0.96	19.2
Operation 3	5	0.96	4.8
Operation 4	120	0.24	28.8
Operation 5	30	0.72	21.6
Total (average cycle time)			

SIMULATION

Follow a body of imaginary cases (A, B, C etc) through the process, allowing for queueing, variations of timing etc

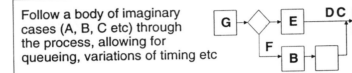

Analysis: 97% took <5 days.

PROTOTYPING

Input screen for Operation 1 like this?

or maybe this?

try out these and other variants

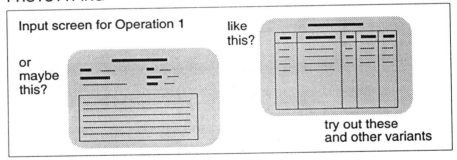

PILOT SYSTEM

Implementing this fully-designed new process in one part of the organisation - with live cases; not a test

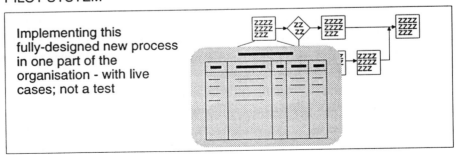

tive information is added (how much work each operation requires etc), **calculations** can be made of the process performance (what the average cycle-time of a case will be etc). This calculation is best seen as a separate activity from descriptive modelling; true, the two may be closely linked (eg handled by the same piece of software), but the link may also be loose (eg modelling by one piece of software and calculation separately in a spreadsheet). Descriptive model and calculation need to be kept consistent, but not necessarily at the same level of detail: you might model in considerable detail, but calculate at a more summary level.

● **Simulation** is a more extreme form of quantification. Simulation software can track the progress of a large body of imaginary cases through the process. This should generate detailed and interesting figures (eg 97% of cases were dealt with in less than five days). This is like studying the performance of an aircraft with calculations to simulate what happens when it is subjected to a variety of conditions. The simulation is abstract; it doesn't mean that any physical aircraft or re-engineered process yet exists.

● At some point an aircraft engineer builds a physical version of the designed aircraft, even if only in wood. This allows people to get a feel for the design that is not possible with an abstract diagram, however detailed. Similarly, during re-engineering design, you may set up the software for a crude, simplified **prototype** version of the new process, that can be used and tried out by team members and perhaps a wider audience.

● A **pilot**, on the other hand, is normally a phase of implementation, not design. It is like the first real aircraft manufactured of a new model. Though probably not optimal in every respect, the aircraft is certainly not expected to crash and kill people. Similarly the pilot version of the process is expected to handle live cases correctly; it is not a mere trial.

Structuring the Design Approach

Developing a design over time

Linear approach
One complete thing after another in natural sequence

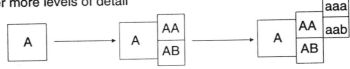

Top-down approach
Adding ever more levels of detail

Iterative approach
Adding ever more detail, and also going back to amend previous design

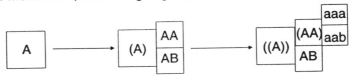

( ) amended version

Distinguishing Kinds of Structure

Various activities may fit together within the larger activity of
process design. As the diagram above suggests impressionistically, there are three kinds of structure to be distinguished:

● **Linear,** ie first decide or design one thing completely, then go
on to the next thing;

● **Top-down,** ie first decide or design one thing, but at a broad-
brush level of detail (level 1), then work in more detail (level 2)

at specific parts of the whole design, but without doing anything that causes any change to level 1; then work at level 3 etc;

● **Iterative**, as top-down but allow, expect and encourage changes at one level to cause changes at other levels, including higher levels.

These distinctions apply to almost any human activities directed towards making something complicated. For example, if you first write the text of a book, then draw some illustrative diagrams and then design the cover, that is a linear approach. The distinction between top-down and iterative is more subtle. If you first outline the structure of a book as (say) ten chapters (level 1) and then write the text of each chapter separately (level 2), that is top-down. An iterative approach would be more like the following: make an outline of ten chapters, then write an incomplete rough draft of most of the chapters, then merge some chapters and split some others, then write some more text, then revise the chapter structure again . . .

Magic Square and Project Structure

Among their principles of redesign, Petrozzo and Stepper include designing the process to handle the simple, common cases (perhaps 80%), rather than creating a robust but complex process for all possible situations. You can always resort, they say, to manual, labour-intensive steps to handle the more complex cases. This broad-brush advice confuses two different things rather dangerously:

● Sometimes designing for the majority of cases only is right. In storing a customer's name in a database it may seem best to decompose each name into (say) title, initials, surname and suffix. It is true that there might one day be a customer called 'His Beatitude Alexius VII, Nestorian Patriarch of Achaea', but to design the database so that this name, as well as any other conceivable name, was decomposed into all its elements would be inordinately complex for little practical benefit.

● But the same argument doesn't apply to all process features. If a new design for a cheque-clearing process has a flaw that causes payments made on February 29 to be accounted for wrongly, you will scarcely agree to live with that, merely because it only arises once every four years. This is a crude example to make the point that some process features by their very nature must be capable of handling any combination of circumstances — even if they are made twice as complex in order to handle the last 1% of unusual possibilities.

What has this to do with the structuring of design-approach activities? Compare a feature where the wayout possibilities can reasonably be ignored with one that has to be as good as watertight in all circumstances. To design the former, a simple linear or top-down approach may be possible; for the latter, iteration, despite the associated risks of anarchy, may be needed to find a design both comprehensive and close to optimal.

Most, though not all, re-engineering endeavours aim to make a process more neat and simple. Experience of systems analysis over the decades shows that a new design that seems, on paper, to be neater than the old may turn out to have irritating minor flaws. Once noticed, the imperfections can often be remedied at a small price in complication — but not always. Occasionally, awkward blemishes, minor but quite unacceptable can only be eradicated by far-reaching modifications. The chance of this is greater, the tighter and more elegant a design is.

The dangerous combination of factors to watch out for is that where watertight features are required *and* the new design is to be much more elegant than the old (eg with 12 main operations instead of 30). As the following diagram suggests, the challenge may be like an infuriating *magic square* problem. You can't ever say that you have *almost* solved this kind of problem. Either you have a complete solution or you haven't. It is worthwhile to judge to what degree any particular re-engineering project has these magic-square characteristics.

Aren't most designs for computer systems like magic-square problems? No, with many applications, you can make a top-level design, tackle each part individually, breaking each down further still, and so on. If the outline design is sound, minor faults in one part can be rectified without repercussions elsewhere. But that

Magic Square Problems

Magic Square Problem

Given these seven numbers, fill in the other eighteen, so that each of the five vertical columns, five horizontal rows, and two main diagonals add up to the same total.

This seems close to solution, but is it? Maybe you will have to go back and change numbers already filled in. That could continue indefinitely.

-	21	10	-	-
9	-	-	26	15
-	8	-	-	-
-	-	-	-	-
-	27	-	-	-

19	21	10	-	25
9	18	12	26	15
28	8	13	24	-
17	6	20	15	-
7	27	22	11	-

Business Process Re-engineering Problem

Redesign this process to achieve the same things much more efficiently.

A perfect redesign solution almost complete. Or is it? Perhaps you will have to go back and. change things already designed.

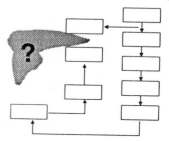

may not be so if an ingenious redesign aims to reduce the old 30 operations to 12 quite different ones.

Nevertheless, re-engineering projects are subject to the magic-square factor to different degrees — and that degree is one variable in decisions about organising a project. The more pronounced the factor, the more carefully the structure of the design work needs to be thought out. In some circumstances the best approach may be a choice and sequence of activities that are far from obvious. In general, the stronger the magic-square factor, the more iteration will be needed around the structure of the design activities.

Design Approach Issues: Overall Chart

If the points just outlined are brought together with a number of others from the *REPRESENTATIVE IDEAS*, five main interrelated issues in design approach can be charted out, as suggested by the following diagram:

- **External influences.** To what extent should you collect and take account of influences, factors and inputs from outside the project team? The main sources are: strategies, plans, policies etc at the level of the whole organisation; benchmarking (ie comparisons with other organisations); and customer surveys. At first glance, the answer may seem simple: exploit these sources to the full. To see the real problem recognise a tradeoff: the more weight allowed to these external influences, the more constrained design discussions will be. If, for example, a survey of customers and competitors doesn't indicate that delivery speed ought to be improved, will you embargo any ideas directed at that aim — or explore them and ignore your survey findings?

- **Designing and proving.** How much of each of these four activities will you do: *modelling, calculation, simulation* and *prototyping*? (*Piloting* is a concern of implementation approach, not design approach.) For each of the first three of these, decisions are needed relative to both the existing and the new process.

- **Impact of IT.** Your approach to design work will be affected by your view of the impact of IT on the design of the particular process concerned. If you see IT merely as second-order detail to be filled in after the main design has been firmed up, that should affect the structure chosen for the design work. If IT is to have a more positive role as a generator of possibilities, the structure should reflect that.

Chart of Five Design Approach Issues

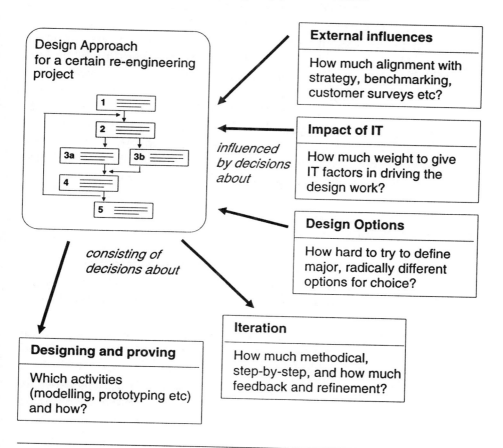

Design Approach for a certain re-engineering project

External influences

How much alignment with strategy, benchmarking, customer surveys etc?

influenced by decisions about

Impact of IT

How much weight to give IT factors in driving the design work?

Design Options

How hard to try to define major, radically different options for choice?

consisting of decisions about

Designing and proving

Which activities (modelling, prototyping etc) and how?

Iteration

How much methodical, step-by-step, and how much feedback and refinement?

● **Design Options.** How much stress do you want to place on the discovery and explicit comparison of mutually exclusive options — as opposed to methodical progress that gradually narrows down possibilities and builds up detail? This depends partly on the nature of the process and partly on change-management considerations (some people enjoy debating alternatives, while others find it cumbersome and confusing).

● **Iteration**. At one extreme a design approach might consist of 12 activities in sequence, with the rule that once an activity is concluded, its product is frozen and passed on to the next activity, and no subsequent revision is allowed (except perhaps if a large blunder is exposed). At the other extreme, there might be four activities that overlap somewhat, and whose products are revised time and again, with the understanding that no product of any activity is ever final until the whole design project is concluded.

A good plan of approach for the whole design project is based on a sound set of decisions, preferably explicit, on these five issues. They are interrelated and need to be coherent; for that reason, the order of presentation above is arbitrary.

CONNECTIONS

The next four briefings deal with decisions of *design approach*.

Briefing 5 discusses external influences on process design, such as organisation-wide strategy, benchmarking and customer surveys.

Briefing 6 goes into the issues with process modelling, sets out a variety of modelling techniques, and reviews the possibilities for automated support.

Briefing 7 deals with the main techniques for proving a design: quantitative analysis — from straightforward calculation through to sophisticated simulation — and prototyping.

Briefing 8 is about the impact of IT on re-engineering; the three main topics are the problem of allowing for IT considerations in the structure of the design approach, the importance of certain specific technologies, and the constraints of IT infrastructure.

Briefing 9 looks at decisions of *process design*. The stress is on the exposition of plausible design options.

Briefing 10 is about *implementation approach*. One of the main topics is the range of possible decisions about pilots.

5.   External Influences

ISSUES

In certain activities outside its brainstorming chamber the re-engineering design team collects information or accepts influences — about organisation-wide strategies or customer opinions, for example. There are more opportunities for good and bad decisions about these external influences than may appear at first glance.

Strategy Alignment

It seems natural and desirable for the re-engineering team to start from a clean slate (aka blank canvas, clean sheet of paper etc), with freedom to consider any idea at all that seems promising — within the area of the chosen process. Much that is written seems to advocate this approach.

But wait, isn't it also sensible to insist that all process designs in the organisation be aligned with some overarching business policy? Suppose the board of a wholesaler of tobacco and confectionery has already decided that the policy of the business is to compete with its rivals on price. It follows that better quality service, though doubtless desirable, will rank lower than enticing new customers through ingeniously calibrated discount structures. Surely, it may seem, this clear strategy should guide the re-engineering of the order fulfilment process.

If that kind of strategic alignment is mandated, it will act as a constraint. Many ideas may bubble up, but only those furthering the company's main priority will be pursued. In other words, the re-engineering team won't have a clean slate. Some fine opportunities for innovation may be available, but, if they don't relate to the compete-on-price policy, they will never be discovered and worked out. If this seems too strict and if the team is permitted to explore attractive innovations not aligned with organisation-wide strategy, then the question arises: What is the point of having the strategic objectives defined exterior to the project at all?

This is a pretty crucial quandary. Decisions here can affect the structure of the process design work. To start out with certain given objectives, and elaborate them methodically to arrive at a new design is one style; to encourage original, inconsistent, inconveniently overlapping ideas to flourish anywhere in the design activities is quite another style. Such different approaches may produce markedly different results.

The choice between strong alignment with strategy and a pure clean-slate approach could make the difference between a successful and an unsuccessful project, particularly when that notable change-management factor, the personalities of the participants, is recognised. What for some people is well-ordered method is for others arbitrary restriction; what for some is fruitful liberty is for others confusing anarchy.

But perhaps the choice need not be so stark. Rather, it may seem, strategy should influence process design only to a certain degree — a different degree in different situations. This is quite a promising line, but a more subtle one than is usually advocated. It needs more exploration.

Customer Surveys and Benchmarking

The other main external influences on design work are benchmarking (comparisons with other organisations) and customer surveys (inputs of one kind or another from customers or potential customers). Here too there are tradeoffs, and they are analogous to those with strategy alignment.

It might seem that the more of these inputs are taken, the better, because the more attuned to market reality the design work will be. But then, it may reasonably be argued, it is extravagant to devote effort to collecting such material, unless you do in fact intend it to influence your design work. That, if it means anything, means that the new design will concentrate on themes raised by the customer surveys and benchmarking, at the expense of others that are not. In other words, you will accept the constraint of not exploring certain promising ideas, about (say) improving service quality or speeding up certain operations, if they are not supported by the results of customer surveys and benchmarking.

Plainly, as with strategy alignment, the right balance between customer surveys and benchmarking, on the one hand, and the freedom of the design team, on the other, will vary with the situation.

REPRESENTATIVE IDEAS

This section deals with the strategy-alignment issue. To what extent should you strive to align the emerging process design with broader organisation-level strategy defined externally to the project?

Alignment with Overall Strategy

Much of the text of the Hammer and Champy best-seller implies that a re-engineering project should redesign a certain process without constraints on any ingenious ideas that come up. A number of other writers seem to mean this when they use metaphors such as starting from a clean slate. A clean slate, the assumption is, brings freedom to conceive radical innovations.

But a second plausible principle is that re-engineering work should be aligned with prior strategic objectives. Thus, in the approach suggested by Davenport and Short, you start by referring to (or if necessary creating) some business vision for the whole organisation (examples given: being more customer-oriented, or improving product quality, or adopting the best practices of Japanese rivals), and (still at organisation not process level) prioritise certain generic objectives: reduce cost, reduce time, improve product quality, and achieve job enrichment for staff.

Then, for those processes chosen to be re-engineered, the design approach ought to be consistent with the business vision and objectives. For example, any work of describing and measuring an existing process should be intelligently selective. If the overriding strategy objective is to reduce costs, the existing process will be studied with stress on its costs; if to reduce cycle-time, the stress will reflect that.

Suppose that in a publishing company job enrichment is the top-priority objective, and the book-acquisition process is chosen as a promising place to achieve it. Then, it follows, during brainstorming and design, you will concentrate on ideas to make the work of the acquisition editor richer (increasing powers of decision, providing access to more information etc), and you won't devote much attention, if any, to innovations directed at other objectives, such as publishing books faster, or reducing administration costs, or improving service to authors.

That kind of stark example is not found in Davenport and Short, and it is easy to see why. The plausible principle of strategy alignment conflicts awkwardly with the clean-slate mandate. The tension is apparent in a couple of the main examples used by Hammer and Champy, where the redesign of a process certainly doesn't start from a clean slate:

● **Hallmark** starts off by defining its beliefs and values, then translates this into a vision, and then goes on to establish business priorities. Only after that does it start designing processes, presumably constrained by all these prior things.

● **Capital Holding** starts with one of those soppy vision statements, headed 'Caring, Listening, Satisfying . . .' Then it moves on to firm up two broad objectives (in translation: improve service, and make more use of information about customers). Then it identifies processes. Only then does it start redesign work. Presumably the two broad objectives are meant to encourage certain kinds of ideas during design and discourage others. Otherwise, why have them?

Carr et al suggest an approach including activities such as getting consensus on strategic vision; projecting future sources of competitive advantage; making a thorough environmental and competitor analysis; identifying possible 'breakpoints' (eg reduce delivery cycle-time dramatically, cut product price in half); and deciding priorities of the breakpoints. Their book is not very detailed, but presumably the intention is that the actual work of redesigning the process should be strongly influenced by all this prior material.

Collecting More Specific Objectives

The discussion so far has been about alignment with some business strategy existing prior to the re-engineering project. A related notion is to go out in the early stages of the project to collect inputs of a general strategic character.

For example, Davenport's book recommends developing 'process visions' early on. These broad-brush outlines of how the new process might look are to be influenced in large part by organisation-wide strategy factors. The book supports this notion by arguing that, if there is no defined 'strategic context', then bureaucratic or non-value-adding tasks may be spotted and eliminated, but the radical innovations will be missed. And again, the argument is, without a vision it is hard to know how to innovate and what type of improvement to pursue.

▼ This seems vulnerable to the *Trivial / False Fork*. If strategic context and vision are taken to mean something so banal as (say) the view that having a fine order-management process would be pleasant, then this is true but so *trivial* as to be worthless. But if strategic context and vision are taken to mean something quite specific, shrewd and unobvious (eg some subtle balance between product price and service quality, targeted at some carefully discerned segment of the possible market), then the assertion is interesting but *false*, since even without such a context or vision, people often do find innovative ideas. This Trivial/False Fork is one of the most potent techniques available for critical thinking about doctrines of business management. ▲

Prior Quantified Objectives

Deciding early on in the design work to try to cut cycle-time drastically is one thing; setting up the prime test of re-engineering success as a cut in cycle-time to some specific figure is something else.

Davenport and Short describe how GUS Home Shopping re-engineered its logistical processes, with two targets from the start (ie two constraints which presumably determined the approach to design work, and the ideas generated): to receive a delivery from a supplier and within five minutes to have the goods in stock ready for sale to a customer; to make the average cost of a delivery to a customer 60 cents. If this is a fair account of what that company did, specific quantitative targets seem to have been set before anybody could reasonably judge whether they were attainable at an acceptable cost.

Hammer's original article describes the example of Mutual Benefit Life, and here too a quantitative target was set beforehand: a 60% improvement in productivity.

There seems to be something irrational about this kind of prior constraint. Suppose the team at GUS finds a process design that can be quickly and cheaply implemented, has other advantages too, but has an estimated average delivery cost of 65 cents. Or the Mutual Benefit team finds an elegant new design that brings only a forecast 50% improvement in productivity.

If the design teams take their prior targets literally they will discard these possibilities and press on to find other designs that do meet the targets — perhaps requiring an investment twice as large, or having other disadvantages, or missing benefits present in the rejected designs. As the illustration suggests, this goes against the general principle in business decision-making of looking for the best buy. And if the prior goals are not to be taken entirely literally, what is the point of them?

Some authorities answer the previous question by raising soft, change-management considerations. Setting an ambitious, prior, quantitative objective may, from a directly-rational point of view, be a poor way of arriving at the best-buy process design, but it

The Trouble with Prior Targets

Week 1	'The prior target for the new process is productivity improvement of 60%.'
Week 3	'In rough outline, we see three main possible designs: A, B and C ...'
	'A seems too modest to meet our prior target. Concentrate on B and C.'
Week 5	'Having filled in some more detail, we think B might bring an improvement of about 50% and C about 75%.'
	'B is of no interest then. Concentrate on C for the rest of the project.'
But	There is no reason to believe that C is the best-buy, most sensible choice.
	Perhaps A is the only wise option, because both B and C are far too risky.
	Or perhaps a careful financial appraisal of productivity gain, required investment and timescales would show that B gave a better return.
	If you follow a prior target slavishly, such questions are never even raised.

may still be sensible if it stimulates and excites and drives designers to stretch their imaginative talents to the utmost. They may then produce innovations that would never have emerged in a coldly rational environment.

There is certainly some sense in this view, but it is usually presented much too coarsely. Though ambitious objectives will exert soft, change-management influences, their force will vary considerably from one situation to another. Moreover, the influence may be negative just as well as positive: a design team astute enough to see that the targets are absurd may easily become cynical and demotivated.

DISCUSSION

This section comments on the representative ideas about strategy alignment within the design task, and then goes on to develop some analysis.

Strategy Alignment: the Essential Dilemma

There seems to be an impossible dilemma. On the one hand it seems undesirable to hobble the creativity of the design team with prior constraints. On the other, it seems wrong to grant a design team freedom to pursue aims divorced from the organisation's overall strategy.

The way of defusing the difficulty is to notice that anyone who strongly favours one side or the other of this argument is probably allowing an unstated assumption: that there is one standard way of carrying out any re-engineering project. With that premise there is no satisfactory, universal way of resolving the above dilemma. But, once the instinct to hold one invariable doctrine on the subject is overcome, a more viable position is open:

● Every re-engineering project needs to be based on some attitude towards the question of strategy alignment.

● In some circumstances the most astute attitude is: 'start from a clean slate, and come up with whatever you think best for this process'; but in others: 'redesign this process, concentrating on ways to cut cycle-time, without increasing running costs.'

● As the following diagram suggests, there are tradeoffs between slim and substantial definitions of strategy alignment. Moreover, this is not an either/or choice; many gradations of definition are possible. What is best depends on the circumstances. The attitude to take on strategy alignment is one of the main generic decisions of design approach.

Strategy Alignment and Process Re-engineering

A slim statement of strategy alignment

'We aim to be an upmarket supplier of goods; therefore
we want a high-quality order fulfilment process.'

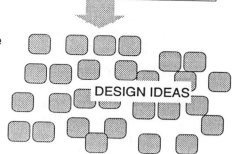

Advantage:
allows a great many bright ideas to be considered; some could be winners.

Disadvantage:
resulting process design may not be perfectly consistent with objectives and activities elsewhere in the business.

A substantial statement of strategy alignment

'To fit our broad business strategy,
we want an order fulfilment process,
that reduces cycle-time a lot,
and need not cut costs particularly,
and can be live within nine months,
and allows for a doubling in product range
and brings in keener discount structures
and . . .
and . . .'

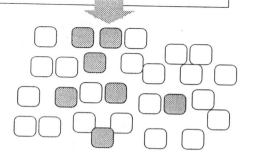

Advantages:
resulting process design is likely to be consistent with objectives and activities elsewhere in the business; also, fewer options mean less design work.

Disadvantage:
some splendid possibilities may never be noticed and discussed.

Strategy Alignment: Examples of Complications

With all processes, but particularly those other than the transaction-based kind, the relation between strategy and process design may be less simple than this briefing has so far suggested. The building-permit process in the town hall of Rheden is an instructive example. It may seem that the natural objective, consonant with organisation-wide strategy, is to approve or reject applications with faster throughput and fewer staff. But there are some complications with this straightforward-efficiency policy.

● At present, substantial effort is spent in helping applicants whose plans are unsatisfactory to revise them with a better chance of success. Slashing this helpful service is one easy way to improve efficiency. But should that be the policy? Maybe the drive for straightforward-efficiency should be qualified by a rider that the present level of helpful service should be preserved, no more and no less. But why? The service that prevails at the moment is an arbitrary one. Should not the issue be decided more actively?

● Moreover, amount of helpful service is a crude concept. What about effective targeting of service? You might cut out certain work revising plans that the applicant's own staff could do almost as well, and use the resources saved for something more valuable. Perhaps improvement in that direction should be part of the policy — though definition and measurement is problematical.

● Since the process is not entirely mechanical (eg aesthetic judgements may come in or judgements about the reliability of long-term demographic and economic forecasts), its outcomes (whether to accept or reject an application) can be of good or bad quality. Should raising the quality of outcomes be an objective, even at the expense of straightforward-efficiency?

● In considering all the above there is an important feedback to consider. After a rejection, the applicant may invoke appeal procedures, perhaps so costly and time-consuming for the town-hall staff that they wipe out the gains of straightforward-efficiency. Perhaps time spent commiserating patiently with an applicant whose application has been rejected is a good invest-ment if it makes an appeal less likely.

Some of these considerations come together in examining the single most striking measure of efficiency: average cycle-time. In a matter-based process like this it can be surprisingly difficult to define cycle-time clearly. According to Dur's book, the Rheden people considered three definitions:

● *Either* (as at present) cycle-time is the elapsed time from application through to either definitive decision or voluntary withdrawal of the application (eg if, through discussions at the town hall, the applicant sees that there is no chance of a successful outcome). This means that time spent by the applicant on revisions or clarifications along the way will be part of the cycle-time.

● *Or* as above, except that any elapsed time when the case is back with the applicant for revision is subtracted from the calculation of cycle-time.

● *Or* cycle-time is the time from application through to definitive decision, or definitive voluntary withdrawal of the application, *or the first revision, however minor.* As soon as a revision is made to an application, the case is closed (as far as cycle-time statistics are concerned), and the cycle-time clock starts over again on a new case.

Manifestly, choice between these definitions is bound up with policy on helping applicants modify their submissions — perhaps the most tricky issue in the re-engineering of this particular process.

The Telstra complaints-handling process described by Ben-dall-Harris raises similar talking-points. For example:

● It seems a sound investment to spend some time after a complaint is resolved in phoning to check that the customer is content and fully understands the explanation or compensation. But how far in this direction should you go? How can you measure whether you are spending the right amount of effort on this kind of activity in the right way?

● How can you set targets for the complaints-handling process as a whole? Suppose it was true, as Telstra believed, that the current level of complaints was rather low, because many customers thought there was no point in complaining or felt intimidated by surly complaints clerks. If so, presumably a better complaints process should result in more complaints. But this seems paradoxical.

What is the conclusion from all these complexities? At the very least, that, for certain processes, delicate debates about the definition of overall objectives may be required.

REPRESENTATIVE IDEAS

Some authorities recommend use of extensive, detailed customer surveys and industry-best benchmarking as inputs to design work. How feasible and generally applicable are such methods?

Varieties of Benchmarking

Everybody agrees that benchmarking can be worthwhile. But there are several forms of benchmarking: getting clever ideas from other organisations with successful processes is one thing; collecting quantitative data on competitors' processing times or even costs is another.

Hammer and Champy concentrate on collecting ideas rather than quantitative data, and warn of stifling creativity by merely copying what others do. Davenport's book too concentrates on the ideas, talking of 'innovation benchmarking'. He points out sensibly that you need to study the other organisations at first hand, since third-party accounts (such as his own book and all other books and articles) may well gloss over material factors.

Cross et al advocate the collection of both ideas and quantitative data through benchmarking. This is all very well, but there is hardly any practical advice, about *how* to go about what may be an immensely difficult task. (Admittedly, they do mention the 'International Benchmarking Clearing House' in Houston, Texas.) Are you going to ask your closest competitor for detailed cost breakdowns? Will you sue if it turns out later that you were given false data that caused you to re-engineer a process to meet unrealistic or unnecessary objectives?

Petrozzo and Stepper are in favour of benchmarking to spark ideas about generic processes found in many industries, such as order fulfilment. They raise the natural question of why other companies should be prepared to give you such precious information; their answer is that the others will be after valuable ideas from you in exchange. These authors also recommend collecting quantitative data on competitors such as unit costs, customer satisfaction and delivery intervals. They acknowledge that competitors are unlikely to impart such sensitive information, but say that it is often available (at a price) from industry consultants. The obstacles indicated by Petrozzo and Stepper are incontrovertible, but the solutions they offered seem threadbare.

Customer Surveys

Nobody is against justice or liberty or customer satisfaction or customer focus. Hammer and Champy (on Taco Bell) say that the customer must be the starting point for everything in re-engineering. Of course, but the interesting issue is: *How much* and *what kind* of study of customer opinions and needs should you do?

Carr et al give the example of PHH FleetAmerica, which asked customers to rank the company and its competitors on 89 different attributes. In addition, focus groups and advisory panels were set up. One drawback is pointed out: only the most enlightened customers may have the vision to suggest worthwhile innovations. Market research at Hewlett-Packard is said to have shown once that an electronic calculator would have no market, since everyone was content with the slide rule.

Davenport judges that questionnaires sent to hundreds of customers (as at PHH FleetAmerica, presumably) are of little use — at any rate as generators of imaginative ideas. Deep, unstructured discussions with a few selected customers are a better approach. Even so, his experience is that customers usually suggest straightforward objectives, eg cut cycle-times, rather than surprising innovations.

Cross et al describe an elaborate procedure for studying customer needs, but this is outside their main methodology for re-engineering. The book makes some sound points: complaint analysis may give data too negative to stimulate innovations; surveys give averages, and thus submerge the views of the shrewdest customers; moreover, surveys collect only data that is expressible in statistics. Another good insight is that the study of customers may suggest new, unobvious ways of dividing them into categories with different needs, and this may have implications for process design.

▼ What does it mean to say that the customer must be the starting point for re-engineering? If it means that you should only re-engineer processes where the customer is directly involved, it is *false*, since that would rule out Ford's procurement process, the most famous of them all. If it means that you should concentrate on things that will make you a better organisation (eg improving procurement), and thus, directly or indirectly, lead to better quality, service or price for the customer, then it is *trivial*: just another way of urging that the business be managed well, as opposed to badly. Many other assertions about management, particularly quality management, are vulnerable to similar criticism. ▲

DISCUSSION

If the work of collecting benchmarking and customer-survey data cost nothing it would probably be worth doing. But in fact it has a price, and it may not be a bargain. The issue then is whether effort expended in benchmarking and customer-survey work is a sound investment.

Distinctions between Benchmarking Activities

A good start is to differentiate the kinds of activity that may be called benchmarking. None of the authors above give an adequate breakdown. There are two dimensions to consider. *Whom* do you look at? *What* do you look at?

The four main answers to the first question are:

- some other closely-comparable unit within your **own organisation,** or perhaps a joint-venture or strategic-alliance partner;
- a **competitor,** or company in a closely-comparable industry, or in the same industry, but a different geographic market;
- a company in a **completely different** industry;
- your industry **as a whole,** or the whole range of the economy — as opposed to any specific company.

The three main kinds of information that may be sought on any of these four dimensions are:

- **quantitative** data about costs, timings etc;
- **factual** but non-quantitative data, eg guarantee conditions or policy on return of unwanted goods;
- **ideas** for doing things: processing an order, enriching a clerical task etc.

Putting together these two pieces of analysis gives a 4x3 matrix, or 12 kinds of benchmarking activity — some easier to arrange and more fruitful than others. This generic catalogue of benchmarking possibilities can make it easier to see and decide between the most relevant options in any particular situation.

Benchmarking and Customer Surveys: Continuous Decisions

Whether effort spent on benchmarking work (and customer-survey work; this rider will be dropped from here on) is worthwhile depends on the situation. But the issue possesses a certain subtlety:

- It probably isn't wise to make a single decision at the start to commit a certain percentage of the design team's budget to benchmarking work.
- You may have to do some benchmarking work before you know whether it will yield helpful insights or not.
- If a moderate amount of benchmarking proves worthwhile, then probably it should be pursued. If not, not.

● Thus, though the authors cited above hardly suggest this, it is often wise to treat the decision about benchmarking as a *continuous* decision, affected by ongoing results (like investment in a research programme) rather than a *one-off* decision (like building a new oil refinery).

There are several variables affecting continuous decisions about investment of effort in benchmarking work:

● The work may generate a variety of inventive, though overlapping and perhaps contradictory ideas — richer than those generated by any other activity. On the other hand, it may not: the ideas may be obvious ones that you already had, or could easily have found in a much cheaper brainstorming session.

● The collection of reliable quantitative data may prove very difficult, and what you do collect may be based on dubiously representative samples and large simplifying assumptions. Or, on the other hand, you may discover some rich vein of data that seems worth extensive mining.

● Even if reliable quantitative data is found it may not be very surprising, nor have any great implications for process design. On the other hand, it may generate some design options. If the average cycle-time of the market-leader in your industry is x days, you may outline (say) three design options: option A, with average cycle-time somewhat greater than x, but with some qualitative advantage that (arguably) counts for more; option B, with cycle-time x; option C, with average cycle-time shorter than x, though requiring a very large investment in the new process.

As with most issues of design approach, the challenge here is to decide how to generate and choose between situation-relevant design options most effectively.

CONNECTIONS

There is a close, perhaps even awkward, connection between taking strategic objectives as input to the design work itself (discussed in this briefing), and referring to organisation-level strategy in deciding what re-engineering project with what scope to set up in the first place. Scope decisions are dissected in Briefing 3.

If external influences are given substantial weight early in the design work, then the other activities, covered in Briefings 6-8, may be strongly affected. For instance, if, in consequence of external influences, improvement in cycle-time is regarded as less important than enhanced service quality, this should probably affect the detail and method of process modelling.

6. Process Modelling

Re-engineering usually calls for some process modelling. There are two main clusters of issues for decision. *In principle*, issues concerning the kind of model (how detailed, what graphic conventions, what features of the process to stress etc) are distinct from the choice of technology to support the modelling. You could, after all, carve the boxes and arrows of a process model on a piece of sandalwood. In practice though, the two clusters of issues interact: detailed modelling is too laborious without automated facilities; moreover, some (though not all) software products support specific modelling conventions.

Kinds of Model

There are two main dimensions to decisions about the *kind of modelling* to be done on a particular re-engineering project:

● What degree of detail should the model of the current process have, and how detailed should the modelling of design options for the new process be?

● Whatever the detail, what sort of modelling should be used? Should the model have symbols for different generic types of operation, or highlight the roles of each actor, or stress the data used in each operation, or be in some other style?

Decisions on both these dimensions ought to suit the particular situation. They should also be compatible with each other; it is little use making a shrewd choice of the most advantageous modelling conventions for the particular process, unless the decision about degree of detail is also sound.

There is usually one obvious tradeoff: the benefits brought by extensive detail or elaborate, sophisticated modelling conventions have to be weighed against the costs and effort involved. This is bound up with another issue. A process has many facets (types of operation within it, people's roles etc), and even the most elaborate model cannot do justice to every facet. A decision has to be taken on which facets are most material in the particular situation.

Automated Support for Modelling

A process model is bound to need frequent revision during the design activities, in order to correct errors, record fresh nuances, or bring in new ideas. To draw original diagrams and their revisions by hand is far too onerous. Thus the support of PC-based software is usually essential.

It may seem at first glance that there is no decision of any great moment to be taken here: surely you should decide about the issues of degree of detail and kind of modelling first; after that, you can select whatever software package best meets those requirements.

But things are not that simple. There may be a software package available that is specifically dedicated to the particular modelling conventions you want to use — or there may not. But even if there is, you may still prefer to use a general-purpose, diagram-drawing package. Or you may consider 'repository-based' software that can maintain two models of the same process, each using different conventions, and thus providing different insights. But, as so often with technology, the more advanced products may well be more tricky to use effectively.

There are currently several dozen software products in contention (and very likely more in the near future), and it would be pointless to set out the tradeoff factors for each product relative to all others; the thing to do is to identify the main *generic classes of product*. This kind of analysis can show how, in general, products belonging to one class have advantages and disadvantages relative to products in the other main classes. Then it becomes clear that the real decision is about the class of software product most suitable for a given project. If that decision is right, it may scarcely matter much if, through lack of time or knowledge, you choose only the second or third best in the class.

▼ Magazine articles surveying products in a certain field are often written as if the main point were to recommend one or several specific products above all the others available. Such findings are of limited use, since they are out of date within a few months. But *frameworks for classification* that make sense of the offerings on the market age much less rapidly. The real tests of any magazine article are the shrewdness of the product categories it sets out, and of the criteria it signals as material in differentiating products. ▲

REPRESENTATIVE IDEAS

This section concentrates on those issues in process modelling about which a reasonable amount has been published.

Process Modelling: Representative Approach

Cross et al give an account of an approach to detailed process modelling. It is a good representative of the methods of a substantial number (though certainly not all) management consultants on large-scale re-engineering assignments.

With this approach you make detailed models of both the current and new processes, each at the same level of detail. This may mean that, if the new process is more elegant, the model of the current process will be more voluminous than that of the new.

Their modelling conventions are fairly simple. Operation-boxes are linked by three kinds of paths: the major path taken by most cases; a minor path taken by a regular minority of cases; and another minor path for rare, exceptional cases. Different information media, eg electronic link, phone, mail, physical transport, or a document have their own icons. Also, the operations on the model are arranged in several bands, typically (but not necessarily): customer or end user; front line or distribution; back room operation; centralised support or information systems; vendor ie supplier contact.

The operations in the model are also appraised to distinguish value-adding from non-value-adding operations (though the model itself contains no graphic conventions for this). Criteria for the distinction are such things as: Would the customer be willing to pay for the operation? Thus non-value-adding operations are those such as sorting and filing mail, verifying and reviewing other operations, and performing corrective actions in error feed-back paths, *as opposed to* worthwhile tasks such as (presumably, though no clear example is given) printing out and posting off insurance policy documents.

At the end of the whole design activity, if all goes well, impressive claims can be made along the lines: 'Under the old process only 6% of steps added value (16 out of 284). With the new process 33% (28 out of 85) add value.'

▼ The trouble with these statistics is that they depend on vague criteria (such as customer willingness to pay) to distinguish the value-adding operations from the rest. Is it true that the average customer of an insurance company is not willing to pay to have the letter applying for a new policy sorted in the mailroom, nor willing to pay to have the new policy document checked for accuracy, but is willing to pay to have the policy document posted? Surely some customers, perhaps working in an office themselves, understand that mail-sorting and some degree of checking are usually inseparable from a decent service.

This value-adding concept is an example of something found in many parts of management literature: the vaguely plausible, impressionistic concept that is as elusive as a butterfly when you try to express it with any precision; others are core competence or high-quality product or information (as opposed to data). ▲

Process Modelling: the Issue of Detail

To suggest the granularity of analysis Cross et al show part of one model of the underwriting process (ie making of new policies) for an insurance company. It has 284 operations on 13 pages. This may seem a lot but, though the authors don't say this, it may give a misleadingly trim impression. First, it excludes procedures for the amendment of policies (which can be more complicated, with recalculation of premiums and with only certain types of change being allowed under certain conditions). Also, the example company has only one type of policy; if there were motor and household and personal accident and loss-of-profits and product liability and twenty other types of policy, a model at the same granularity would need many more than 284 operations.

Indicating the proposed granularity of any technique is essential. Morris and Brandon recommend a certain modelling method, but whether it is meant for a 20-operation or 2000-operation model is unclear.

A large body of opinion holds that modelling such as that of Cross et al is too detailed in most instances, and examples of companies stuck in quicksands of model detail, often under the influence of consultants, are mentioned in several articles. Hammer and Champy warn against the danger of analysing the present process in great detail. They say that you only need a high-level view to support the intuition and insight needed for a new design. Of course, this piece of advice can be interpreted in many ways.

Davenport puts a view that is logical, given its premises. Accept the principle that the design should develop in a top-down way, from decisions of organisation-wide strategy, leading to broad characteristics of a new process, then expanded in more detail, and so on. From this it follows that:

● The effort devoted to studying the **existing process**, and the aspects to concentrate on, should depend on the broad picture of objectives and outlines for the new process already built up. If it is already clear that certain whole pieces of the existing process will have to be thrown away, there is little point in modelling them in great detail.

● Similar factors should determine the effort devoted to modelling the **new process**, and the aspects to concentrate on. Suppose the main target for the new process is improving service quality, as opposed to making quantitative gains such as reduced cycle-time. Then it will be pointless to encrust the model with subtle details whose only point is to permit meticulous studies of cycle-time.

Process Modelling: Representative Techniques

The books by Andrews and Stalick, and by Johansson et al sketch a variety of modelling techniques, but without giving much account of their suitability for different purposes. Various other authorities advocate one particular set of modelling conventions and imply — albeit with varying degrees of boldness — that this one method is effective in the great majority of situations likely to be encountered:

● **RAD** (role activity diagram) modelling (well described by Huckvale and Ould in the book edited by Spurr et al) gives great attention to the *roles* of the actors in a process. A process such as reimbursing employees' expenses may consist of (say) 20 related operations altogether, but it is illuminating to distinguish several role-players (expense claimant, accounts clerk, chief accountant etc), each carrying out certain operations — either alone or interacting with others. The model is drawn so that, above all, operations belonging within each role are kept clearly separate.

● **IDEF0** (described briefly in the book by Johansson et al) is a method adopted as standard in some parts of the US Government. Each box represents an operation. One or more arrowed lines enter a box from the *left*, corresponding to inputs: things (data or physical objects) consumed or transformed by the operation. Lines leaving from the *right* of the box correspond to outputs: things (data or physical objects) produced by the operation. Lines coming in at the *top* are for controls (policies, rules, conditions, laws, standards etc), that determine the way the operation works. Lines coming in from the *bottom* are for mechanisms or resources (people, machinery etc) used to perform the operation.

● The **Winograd/Flores** 'Conversation for Action Model' (summarised neutrally in the *Economist* article of 11 December 1993) is based on the following bold generalisations. Every piece of work has two people involved: the performer (eg expense claimant) and customer (eg accounts clerk for whom the expenses form is filled in). Also, every piece of work can be decomposed into four pieces: preparation, negotiation, performance and acceptance. The contention is that *any* business process can be broken down into a hierarchy of two-people, four-phase items — with some practice, of course. This approach fits some processes very naturally (eg buying a hamburger from a stall), but with others (eg month-end expenses procedures, with a variety of people required to authorise different kinds of expenses) more ingenuity is needed.

▼ These representative examples suggest questions about the *status* of any specific modelling method. Is the method claimed by its advocate as generally superior to all others (at least for the great majority of processes)? If yes, what good reasons are there to support that claim? What is inferior about all the others? If the method is not claimed as *generally* superior, then, if worth using at all, it must at least be particularly suitable for certain kinds of processes rather than others. Where are the guidelines to help anyone decide whether any particular process is of the kind that the method fits?

Most advocates of one specific method seem to assume or hope that such questions won't occur to their audience. A similar line of enquiry can be applied to standard methodologies in any area of management. ▲

DISCUSSION

This section concentrates on surveying the main varieties of modelling available for choice.

Distinguishing Three Types of Modelling

At the outset, there is a basic point to clarify. Many modelling conventions exist. Some differ only in trivial ways, representing much the same content, but in different graphic formats; choice between these is little more than a question of taste. But others differ so much that it is fair to talk of different *types* of modelling or of models stressing different *facets* of a process. Here choice of modelling method can have far-reaching implications.

To begin the untangling, distinguish three broad *types of model* that can help in process design:

• Simple aids to clear thinking, notably the **hierarchy diagram** and the **matrix diagram**. A hierarchy diagram might show (say) a bank's 'loan placements' process to consist of four sub-processes, one of them 'identify assets for resale'; this in turn consists of several sub-sub-processes, including 'request credit approval'; this in turn consists of . . and so on. On a matrix diagram, by contrast, each of the sub-sub-processes in loan placement might be given a row and each person involved in loan placements a column. Crosses in the cells of this matrix would show who were involved in which activities. Neither of these diagrams is at all apt for showing what operations happen in what sequence, under what conditions.

• **Data models**, notably the ER (entity-relationship) diagram. These can show, in quite a rigorous way, what data items are associated with a process; eg the entity 'loan' has attributes such as 'agreement date', 'due date', 'borrower', 'interest rate' etc. However detailed, this kind of model is unlikely to give a clear picture of a process, because it does not adress operations, their sequence and their conditions.

• **Process models** are by far the most important to re-engineering. There are many variants, but all attempt to capture the relationships between the operations of a process: that is, sequence, conditions and perhaps other facets too.

Kinds of Process, Modelling Approach

There is a natural tendency for a re-engineering project to concentrate on process modelling. Data modelling doesn't capture step-by-step operations within a process, and, in any case, may be too detailed to be a good investment of effort. It may, however, come in later, after the main design decisions have been taken and the work of developing software and databases begins.

This may sound decisive but, in fact, is only so for *transaction-based processes*. For *facility-based processes*, such as IBM's Latin America information system, the situation is reversed: modelling of data that may be accessed is exactly what is needed, while detailed process modelling is scarcely possible.

With matter-based and project-based processes the considerations are less clear-cut and more situation-specific. Most *matter-based processes* call for process rather than data modelling, though there can be situations where some data modelling during re-engineering may be needed. In any *project-based process* concerned with product development, where engineering designs have to fit together, data modelling is the more useful, though some outline process modelling may be needed too.

On the question of modelling detail there is one safe generalisation to make: authorities such as Cross et al who give the impression that much the same degree of detail (whatever degree it may be) is appropriate for most processes are wrong. Situations are far too variable. Determining the degree of detail is a serious decision of design approach. It is surely misguided to adopt some standard answer and apply it on project after project. However, if processes are distinguished by kind, some cautious generalisations about options and tradeoffs are possible:

● With a **transaction-based process** it will normally be *feasible* to perform quite detailed process modelling. Whether it is also *desirable* is another question. If large quantitative improvements are intended (eg 'the new process will cut cycle-time by 50% . . .'), then detailed modelling may be essential as a starting-point for quantitative study. But if the re-engineering has other objectives, the balance between effort and benefit may be different.

● With **matter-based** and some **project-based processes**, the decision on modelling detail can be delicate. To streamline and rationalise the process (ie make it more like an efficient transaction-processing machine), detailed process modelling may be essential. But if the aim is to enable actors in the process to interact with each other more richly, taking shortcuts or perhaps looping back for extra conferring, based on individual judgement of a matter, then modelling in detail may not be worthwhile. Here the notion of one well-defined form of the process begins to be undermined, and the process takes on more of a facility-based character.

• With an extreme **facility-based process** (such as information to support decision-making at Bow Valley Industries) detailed process modelling may not be a credible option, since there are too many possible paths and feedbacks; a lattice-like diagram with 37 operation-boxes, each with arrows going to (on average) 25 of the other boxes is scarcely worth having.

Challenges in Modelling: an Example

Another factor in decisions about modelling is the demands placed on those who have to do it — in terms of correctness, and of choice of valuable detail. Matter-based processes may be particularly tricky. For example, take a local-government process for handling building-permits:

• The model may need multi-valued branching; eg out of one operation-box the possible courses may be: refuse, accept, postpone for more study, call applicant for clarification . . .

• There may be many ways for a permit-application to fail, each with its own feedback path. You can model every conceivable eventuality, but that is probably over-complicated and will obscure the main flow of the process. But if you are selective, delicate judgements are needed on which paths deserve to be represented and which do not.

• Some decisions within the process (eg to reject or perhaps to seek further advice and go round extra loops) may be based on hard, definable criteria (eg 'Has the applicant filled in the correct form?') and others on soft criteria (eg 'Will the new building spoil the landscape?'), but others may be half way between — based in part on intuition (eg 'Does the cost-estimate seem plausible?'). How far should you go in firming up such criteria in the model?

• Lags, queues and the like may be complicated: refer to the legal expert, available once every two weeks; or postpone decision until some other plan (for the whole neighbourhood), is ready; or even

Process Breakdown - Three Levels of Detail

```
┌─────────────────────────────────────────────────┐
│                                                  │
│          ┌──────────────────────┐                │
│          │ process customer     │                │
│          │ cheque               │                │
│          └──────────────────────┘                │
│   ┌──────────────────────────────────────────────┐
│   │  Breakdown of 'process customer cheque'      │
│   │                                              │
│   │  ┌──────────────┐  not OK  ┌──────────────┐  │
│   │  │ verify cheque │ ───────→ │ return cheque │  │
│   │  │ properly filled in│     │ to customer  │  │
│   │  └──────────────┘          └──────────────┘  │
│   │         │ OK                                 │
│   │  ┌──────────────┐  not OK  ┌──────────────┐  │
└───│  │ examine customer│ ─────→ │ process bounced│ │
    │  │ account balance │        │ cheque       │  │
    │  └──────────────┘          └──────────────┘  │
    │         │ OK                                 │
    │  ┌──────────────┐                            │
    │  │ carry out    │   ┌─────────────────────────────┐
    │  │ transaction  │   │ Breakdown of 'carry out transaction'│
    │  └──────────────┘   │                             │
    └─────────────────────│   ┌──────────────────┐      │
                          │   │ update customer  │      │
                          │   │ account balance  │      │
                          │   └──────────────────┘      │
                          │            │                │
                          │   ┌──────────────────┐      │
                          │   │ transfer money   │      │
                          │   │ to recipient     │      │
                          │   └──────────────────┘      │
                          └─────────────────────────────┘
```

(an awful one) check for compatibility with any other applications also currently on their way through the process.

Facets in Process Modelling

A model of a business process is like a geographical map. Even a very detailed map can't depict everything that might be of interest. It is impractical to show the boundaries of all political units (provinces, local authorities etc), and all roads (motorways down

Data Flow Modelling - a Representative Fragment

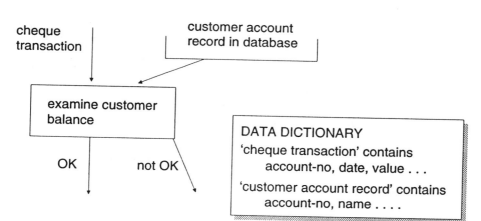

to cyclepaths with all intermediate gradations), and detailed physical relief, and climate zones, and population density, and so on, all together on one map. Any one map concentrates on certain facets and gives less or no attention to other possible facets. Different maps suit different needs: planning a motoring holiday, reorganising local government boundaries, and so on.

Any method for process modelling is really a suggested way of drawing a map to stress certain facets of a process, and not others. This raises the question: What are the main facets of a process worth mapping? Here are the four most common:

● **Breakdown.** As the first diagram shows, it is possible both to represent the flow within a process, and to show how operations break down into smaller operations, at several levels of detail. In all other respects this example is a plain-vanilla model.

● **Data.** The plain-vanilla example implies that certain data must be accessed and used (eg account balance), but data is not represented explicitly in the model. If many operations put data into a database, amend data or refer to it, then it may be desirable to record this in detail. The *data flow diagram* is a long-established form for relating data to process. As the second diagram suggests,

Generic Operation-types - Possible Conventions

Either
put icons NEXT TO
descriptive text

Or, if you prefer a different style,
use boxes to CONTAIN the
descriptive text

Capture
eg key in the data of the new insurance policy

Process
eg calculate insurance tax payable on premium

Duplicate
eg make copy of policy for agent

Control
eg check postcode is valid for the address given

Transport
eg post cover confirmation slip to the insured

Store
eg add new policy to customer file folder

each chunk of data (eg a transaction or a database record) is described quite rigorously in a substantial data dictionary. A much lighter approach is to show just the *things* that hold data, without detail about their content; examples of generic data-holders, for which appropriate symbols can be used, are document (eg a cheque), database, phone message, telex or fax.

● **Operation-type.** In the diagram of the plain-vanilla model each operation has a simple box. Another possibility is to use a graphic representation for each of several types of operation. Many generic operations and graphic forms can be imagined. The example takes a six-way analysis of operation-types offered by Johansson et al, and, spurning that book's taste in graphics, suggests two possible methods of representation.

Modelling the Actor Facet

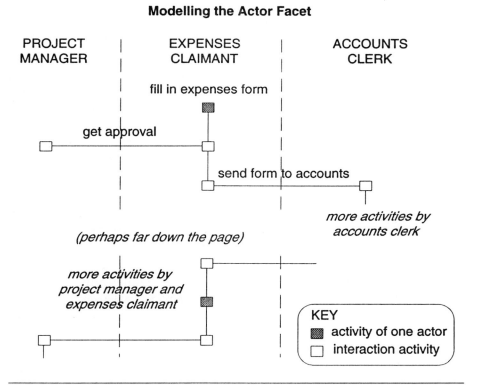

● **Actor**. The plain-vanilla model shows what is done, but not who does it. Some methods map this facet by dividing every operation-box into two parts, corresponding to action and actor. A more radical approach (as in the example — a simplified version of RAD modelling), is to organise the whole format of the model primarily by actor. If there are many actors, the interactions between them may produce long horizontal lines across the page, and thus become very prominent. This may be advantageous since interactions often bring delays, and therefore deserve scrutiny. Mapping of this facet can hold many nuances; eg two roles — such as purchaser and inspector of goods — can be played by one person, or optionally by two different people, or compulsorily by two people, and so on.

More Facets in Process Modelling

There are more facets of a process that, if studied and mapped with care, can lead to considerable complexity:

● **Rules.** In the plain-vanilla model of the first diagram rules are implied rather than explicit; eg certain rules are used to decide whether or not a cheque is correctly filled in. In another modelling technique rules may be represented much more prominently; eg with a standard syntax for rule-definition using 'if', 'and', 'not' etc; or with different graphics for types of rule (validity-test, importance-test, completion-test etc).

● **Flow-type.** In the plain-vanilla model arrowed lines show the flow of work, and that is all. But if several forms of lines and arrows are used, some varieties of flow can be represented: *usual* flow; usual for a certain *easily defined minority* of (non-error) cases, eg all above a certain value; awkward-to-define *exception* cases, eg clerk feels uneasy about the cheque; and *error* case. 'Batching' is another subtlety. Suppose one operation is 'sort out all incoming cheques'. That box on the model applies to a whole batch of cheques, and is actioned perhaps once a day. But the box verifying that a cheque is properly filled in applies to one cheque and happens once for every single cheque. That may seem obvious, but in some processes it could be a subtlety of flow well worth distinguishing by a graphic convention. If so, two signs are needed: one for splitting a batch into discrete items, and one for merging discrete items into a batch.

● **Queue-type.** The plain-vanilla model is silent about what happens *between* boxes. But in practice cases must sometimes wait in queues between operations. Quite a variety of queue-types can be distinguished and modelled: first-in, first-out; priority-affected; temporal (eg wait till end of month); and many variants.

● **Other facets.** As with the map of a country, the list of facets that might conceivably be represented is unending. For example, Cross et al organise their model diagrams in five bands from customer-contact at the top through to supplier-contact at the bottom. This might be called a *customer-proximity* facet.

It is not realistic to expect to do justice in one model to all the above facets. The model would be too complicated to be manageable, and, in any case, certain facets tend to mix badly. If there are (say) four or five levels of breakdown, the model can scarcely stress the actor facet as well, since the role of any given actor may be diffused over several levels.

Facets and Decisions

The survey above helps pinpoint the essential characteristics of any particular modelling method and relate it to other methods. For example:

● The **Cross et al** method is a kind of modest mixture. It provides some analysis of the flow-type facet, but none of the other generic facets (operation-type, breakdown, actor etc); it also has its own facet of customer-proximity.

● The **RAD** method concentrates on the actor facet, has no breakdown facet, and a mild degree of analysis of the operation-type and flow-type facets.

● The **IDEF0** method is rigidly standardised (ie there are four things to record about every operation). It gives some attention to data and actor, but is weak in modelling rules and conditions, and doesn't distinguish operation-types or flow-types. Because of its graphic conventions a diagram quickly becomes congested. Therefore many levels of breakdown may be modelled. The approach seems best suited to processes where there is a strong hierarchical structure, rather than a linear one, that is full of conditional branches and feedback loops.

● The **Winograd/Flores** method is dominated by the bold general theory that everything that happens is to be analysed into four generic operation-types. It also breaks operations (particularly those of its 'performance' type) down into many levels, each of four more operations. There are reports that the number of these four-item elements in a detailed organisation-wide model can run into millions.

This kind of assessment of modelling methods can help with decisions of design approach. You can take a view of which facets

Deciding about Facets to Model

'This *transaction-based process* contains nothing very tricky (eg features requiring subtle analysis of types of flows or queues), but there is a great deal of detail to model. Therefore the most important point is that our model should be **broken down into many levels.**'

In this *matter-based process*, a very big issue is the possible reshuffling of operations between people with newly defined roles. There are many different ways this may be done. To help us find the design with the most sensibly defined roles our modelling will stress the **actor facet.**'

'Streamlining this *transaction-based process* is essentially a problem in optimising throughput in a system that unavoidably has some quite intricate queues. Concentrate on modelling the **queue-type facet.**'

'In this *project-based process*, the challenge is to allow people to work on different parts of the product design, and fit their results together. Our process modelling will be fairly sketchy, but it must convey some insights into the **actor, flow-type and queue-type facets.**'

seem most relevant to the particular process, and either find a modelling method that is strong on just those facets, or modify a method accordingly, or devise your own. The table shows some plausible decisions.

REPRESENTATIVE IDEAS

Not much of analytical consequence has been published about *automated support* for re-engineering. Davenport restricts himself to a list of ideal features, rather than, for example, showing how different forms of automated support suit different contexts. What follows is an original classification.

DISCUSSION

This section sets out generic options for automated support, and offers tentative advice on which are appropriate in which situations.

Gradations of Automated Support

The main possibilities for automated support of process modelling form a set of gradations, where each is, in some way, more sophisticated than the previous one:

● **Gradation 1.** Use a general-purpose software product capable of drawing any graphics that consist of shapes with text. This kind of software is typically used for many other purposes besides process-model diagrams; eg making slides or overhead transparencies for presentations, or diagrams for a book such as this. Any quantitative data (eg time required for each operation) has to be stored separately in a spreadsheet.

● **Gradation 2.** Use a narrower but still general-purpose 'diagramming' software product, meant for drawing any type of extensive diagram that uses some definable set of conventions repeatedly and consistently. Process models fit this definition (along with diagrams for electrical wiring, organisation charts etc). There are two main advantages over the previous gradation. First, when drawing the model you can keep the repertoire of shapes and line-types for your modelling conventions as a kind of palette on one side of the screen; you pick the shape from the palette instead of drawing it each time. Second, 30 sheets may be needed to print out the whole model in a legible size, but from a logical point of view there is just one diagram; gradation-2 software lets you roam across the whole model on a PC screen (or else, or as well as, up and down through levels of detail).

● **Gradation 3.** Use a software product that is much more specialised: it draws process models based on one particular set of conventions. The advantage, at least in principle, is that the software can check the *content* of the model. For example, if it 'knows' that according to the conventions in use a certain shape should always have one arrowed line going in and two coming out, it can flag an error if that is not so.

● **Gradation 4.** As 3, but the software can also store quantitative data (eg average times, costs and volumes) with the model, and use this data to perform calculations of average throughput time or costs of operations etc. This makes the separate spreadsheet needed for gradations 1-3 unnecessary, unless (which is quite possible) the particular calculations you want are not provided by the gradation-4 software.

● **Gradation 5.** As 4, but the software also uses the 'repository' concept to permit the production of models of the same process in several styles. This concept is so significant that it must be looked at in detail.

The Repository Concept

On some projects it may be worthwhile to model one process in several styles, stressing different facets. Perhaps you feel at home working with diagrams in the conventions suggested by Cross et al, but for some investigations, where allocation of roles is the main facet of interest, you would like to draw RAD diagrams. Using different software to maintain two models would be unacceptably laborious. Besides, it is far too easy for inconsistencies to creep in. The only viable approach is to use one piece of software powerful enough to switch from one style of model to another and keep both consistent.

To see how this can be done, distinguish between the graphic model-diagrams themselves and their underlying data. Even at gradation 1, as well as gradations 2, 3 and 4, the software has to store data that can generate the model diagrams on screen or page whenever required: the data that (say) box 17 is at the end of an arrow that emerges from box 14, that box 17 is to be a certain shape, to contain a certain text etc. But this underlying data is stored by the software behind the scenes; the user of the process simply works with the diagrams themselves.

The Repository Concept

Data items held in the repository about each operation in the process

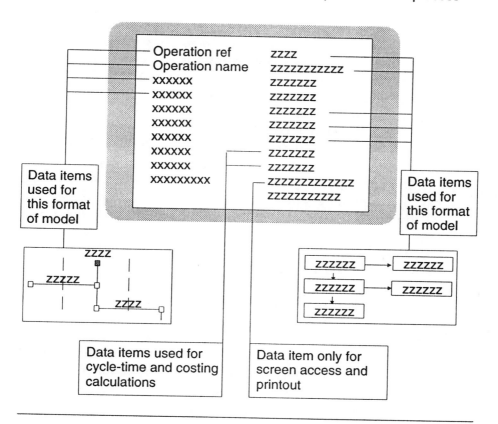

A gradation-5 system stores its underlying data about the process in a careful, flexible way, enabling diagrams to be generated in more than one modelling style. This body of data is said to be held in a *repository*. The diagram gives an impression of the concept.

Any change made to the process design (eg delete a certain operation) can be made just once; the data in the repository is updated; when next any diagram in any style is generated from the repository, it will be up to date. That, in outline, is the solution to maintaining consistency.

The underlying data in the repository is not kept entirely behind the scenes. It can be accessed independently of the diagrams themselves. This allows you, if you are careful, even to alter the modelling approach as you go along through a project. For example, suppose the repository is set up to classify every operation in the process as one out of six types and to differentiate them graphically in a model diagram. After doing some modelling, you decide that types 4, 5 and 6 are not worth distinguishing after all. You can alter the operation-type data within the repository (eg make all types 5 and 6 into 4). This should be easier and less error-prone than amending all the diagrams.

But the repository concept holds nasty snares. As a simple example, suppose you are using two styles of model: style A distinguishes six types of operation (and gives each its own graphic shape), while style B depicts all operations with the same graphic shape. Suppose you are working on a diagram in B format, and you add a new operation. You don't need to say whether it is one type of operation or another (in fact, the model format won't permit you to). But the updated repository is now invalid; it contains an operation whose type is undefined, and therefore can't be shown on any diagram in A-style. Some safeguards have to be built into the repository software to prevent such a thing happening. But this can get very complex. Suppose there were (say) four styles of model, and they differed in quite intricate ways . . .

A minor but useful extension of the repository concept is to store a variety of other pieces of information along with the underlying data for the process model: texts stating the organisation's mission, or analysing its market position, competitive weaknesses etc; names of managers, their titles, responsibilities etc; matrix diagrams (eg showing which departments are concerned with which processes). Even if your system is at one of the other gradations, you can easily store such non-model material on a PC, but a good gradation-5, repository-based system should be more convenient.

▼ The product information that follows is taken mainly from suppliers' marketing material available at the beginning of 1995. Its purpose is to show the *classes* of product that exist, and the *typical issues* that affect choice between one class and another. Thus it provides a framework for relating any particular product to the market as a whole.

The technique of determining the generic material points before appraising any particular product can be applied to any situation where competing products need to be sorted out. In this instance, one or two candidate products (not mentioned below) were described in brochures that — though strong in colour photos of politically-correct groups of people conferring in boardrooms — did not say anything specific enough to determine such fundamentals as the process facets modelled and the gradation of facilities provided. ▲

Representative Software Products: Gradations 1 and 2

Software to support modelling on re-engineering projects is still quite a new product category. Each supplier, understandably enough, produces glossy brochures to stress the virtues of the particular product, without stating the limitations. The method of gradations helps marshal the mass of marketing material in a clear, objective way.

At *gradation 1* there are many well-known products, such as CorelDraw, Freelance Graphics, Harvard Graphics and Persuasion. However, there is little point in saying much about them, since a gradation-2 product will generally be a better buy. It brings the advantages mentioned above and, in the context of practically any re-engineering project, the cost is nugatory.

Diagramming software in the *gradation-2* class is a market with keen competition between products on both price and features. Here, merely to give a flavour, are five:

● **ABC Flowcharter** is specifically for diagrams depicting the flow of a process — unlike (say) diagrams of electrical wiring. This narrower focus helps it to be relatively strong in labour-saving features for chores such as aligning boxes, drawing lines between them, and so on.

• **CorelFlow** is aggressively priced, from one of the major players in graphics software, and is strong in clip-art; you can use tiny drawings of filing cabinets or computer terminals or people sitting at desks as symbols in your modelling conventions. Other products allow this too, but this product's selection is particularly rich.

• **Intellidraw** has a rather idiosyncratic mix of capabilities; eg perspective distortion of text and shapes, or gradient fills of shaded boxes, or controls on letter spacing for typographic refinement.

• **MacFlow**, apart from its intrinsic merits, runs on the Apple Macintosh, which, in some quarters, still has a certain aura of superior elegance: a change-management, rather than directly-rational, factor.

• **Visio** is the diagramming product that most new products in the category tend to get measured against.

Representative Software Products: Gradations 3 and 4

A *gradation-3* product is really a gradation-4 product without the quantitative data. Most in this class are part of larger sets of CASE (computer-aided software engineering) software tools. They are primarily concerned with the fine detail of developing computer systems and databases rather than with process modelling for re-engineering.

• **Bachman/Analyst**, **EasyCASE** and **System Architect** are arbitrary representatives of the class of CASE products that provide gradation-3 support for *data flow diagrams*, with breakdown into levels (practically essential for such detailed modelling). Some products offer a choice between conventions for this kind of model; Yourdon/DeMarco, and Gane and Sarson are the best known.

• **Object Management Workbench** is a rare gradation-3 product for modelling process facets other than data flow. It has fairly simple conventions; they include a breakdown facet, and business rules applying to operations (eg the criteria for whether a cheque is correctly filled in) are held in a standard format away from the model diagram by a separate software tool. The diagram and rules software tools can be integrated with another tool that records data structures.

Software products that provide *gradation-4* support (ie average time, value or cost data can be attached to items in the model) differ considerably in the process facets they address:

● **Action Workflow Analyst** is concerned exclusively with the Winograd/Flores method of modelling. This is sometimes claimed quite fervently to be a simple yet immensely powerful, universal modelling method, but not everybody finds that convincing.

● **BPwin** makes models in the conventions of the IDEF0 standard. It can link up with data definitions held by the companion ERwin product.

● **MAXIM** really lies midway between gradations 3 and 4. You can attach costs, times etc to the model and categorise them, but to be analysed the data must be exported to a spreadsheet. MAXIM's modelling conventions arrange operations in bands corresponding to different departments; this is essentially a variant form of the actor facet. To judge from the published examples, the modelling method has difficulty with operations where members of several departments work together.

● **RADitor** supports RAD diagrams. Thus it concentrates on the actor facet, has no breakdown facet, and has a mild degree of analysis on the operation-type and flow-type facets.

Representative Software Products: Gradation 5

Repository-based products are a fascinating but somewhat immature software category. Products differ in stress:

● **ANSWER:Architect** is a method-independent software backbone, including repository, that can support many kinds of modelling, not just process modelling. For re-engineering, this product is marketed in a form called **ANSWER:Cabe** with its repository already set up to generate several styles of model. There is little stress (at any rate in the marketing documentation) on the user getting at the repository in order to define further styles of model or to alter the styles supplied.

• **Business Design Facility (BDF)** comes with its repository equipped to generate six styles of process model diagram. However, in some instances, the content of one diagram-style is just a summarised version of another. In others, the differences between styles are largely of graphic format; you would normally choose whichever suited your taste and ignore the other. This is a very different thing from having models describing several different *facets* of a process, and thus giving different insights. In fact, scrutinised for its facets, BDF proves to be concerned with breakdown and little else. The documentation mentions the possibility of getting at the repository in order to define further styles of model, but the scope and constraints in doing this are not spelt out. The repository enables the process models to be kept consistent with data models, and in due course, used as a starting-point for software development work.

• **ProcessWise WorkBench** is specifically meant for re-engineering. It comes supplied with its repository set up for just one simple style of modelling. The documentation encourages the user to amend the repository to add extra refinements to this basic approach, or to define other styles of model and other pieces of data to be stored. How how far it is feasible to go in this direction without running into constraints restrictions arising from the need to keep different models of a process consistent is left unclear.

It would be churlish to complain strongly about inadequate documentation of repository products. The products are complex and this is a new market. Suppliers and customers are still discovering the real needs, problems and solutions.

Analysis and Decisions: First Round

As always, software sophistication should match the needs of the situation. Here are some initial pointers that emerge from the account above:

● Although many people do take the gradation-1 approach, **gradation 2** will almost always be superior. Even if you already use a gradation-1 product for other purposes, the marginal cost of a gradation-2 product will probably be outweighed by the marginal benefits.

● A **gradation-4** product is much superior to a gradation-3. The only real argument for a gradation-3 choice would be as part of a larger CASE-based approach, in which the process model has multiple purposes: assisting re-engineering design work, but also serving as the springboard for generating the software and databases of the computer application system that will be needed. But here, even if the approach is successful, it has to be conceded that an inferior means of supporting process design is being chosen for the sake of a quite different consideration, speedy software development.

● **Gradation-5** products pose a special tradeoff problem. The repository concept is attractively elegant. On the other hand, it can raise tricky complications, and the software product must be an advanced one. As yet, the market is relatively immature. On balance, for the time being, it is probably best to choose a straightforward gradation-4 product over a gradation-5, unless you absolutely must have the facility for multiple styles of model — a condition probably applying to only a minority of projects.

● Thus the real choice should usually be between products at gradations 2 and 4.

Analysis and Decisions: Prima Facie Case and Objections

The neatest way to approach the choice between a gradation-2 (general-purpose diagramming) product and a gradation-4 (specific modelling technique, including quantitative data) one is to set out a prima facie case, followed by possible objections. *The prima facie case is that the gradation-4 advantages are decisive:*

● Keeping quantitative data in a spreadsheet separate from the model is unpleasant. You must store not only the timings of all the operations, but also the probability of each operation being reached (eg 3% of cases follow the error path). In a complex process with many paths this can be very awkward to set up, but the real trouble comes when the model is amended. It might take five minutes to alter a few details on the model and an hour to change the logic of the spreadsheet calculations — and even then, subtle errors could creep in. Thus the gradation-4 feature of keeping both descriptive and quantitative data in the same model, and thus automatically consistent, is very attractive indeed.

● The gradation-4 product, that 'understands' the conventions in use, can check for inconsistency and incompleteness, in a way that the gradation-2 product cannot. This point can be compelling in a large model using elaborate conventions.

The question is: Do these generic prima facie considerations hold good in the particular situation you are considering? If they do, the decision is clear; if not, gradation 2 remains an attractive choice. Here are some objections that may apply in some circumstances to overturn the arguments in favour of gradation 4 above:

● Perhaps the gradation-4 product automating the modelling method you want doesn't handle quantitative data in all the ways you need. Maybe it produces some timing and cost figures, but for other pertinent figures, you have to use a spreadsheet. If so, the advantage of holding everything in one model evaporates.

● Perhaps the gradation-4 product isn't as strong as it might be at checking for conformity to the conventions of its specific modelling method. Or perhaps the method itself isn't very elaborate, and so checking for conformity is a rather small consideration.

● The gradation-4 product may automate the specific method you want, but perhaps it is not particularly easy to use. Perhaps, by the standards of the slickest gradation-2 products, that have evolved in a very keen market, the gradation-4 product needs too many mouse movements to draw basic diagrams; or perhaps lining up the content of neat and attractive (as opposed to merely correct) diagrams is far too laborious.

▼ The format of a prima facie case with possible objections, though missing from most business-school books about decision-making, is handy for setting out the arguments in many business situations; eg choosing an advertising agency, appointing a successor managing director, deciding whether a certain new product should be compatible with industry standards, etc. ▲

CONNECTIONS

There is a very close connection between this briefing and the next, Briefing 7, which is about simulation and prototyping. It would be an odd approach first to complete process modelling for a new design in full detail, and only then to begin prototyping or simulating it. The whole point of a test is that the results may not be perfect first time; if so, you will want to go back and model new variants. Thus the activities described in this briefing and in Briefing 7 will often be mingled in an iterative way.

7.   Trying out the Design

Modelling new designs for a process is one thing; testing out and justifying them is another. In a healthy project there will usually be iteration: trying out a possible new design by experiment or by quantification may stimulate ideas that lead to modification of the model, after which the new version has to be tried out ... Even so, the distinction between modelling and trying-out is a handy one for untangling the many issues that call for decisions of design approach.

Activities primarily concerned with trying out a design divide neatly in two: working from an abstract model of the new process design in order to quantify its implications; and experimenting with an actual, but crude and simplified, prototype version of the new process.

Calculation and Simulation

The spectacular slashing of cycle-time (eg from 53 days to five for a product-design process at AT&T) is a frequent theme of re-engineering examples. Improvement on this scale may be possible if there is wanton incompetence to be eradicated, but, the question may be asked, is it credible in a reasonably well-run organisation? Often it can be shown that, even with an apparently sound process, most of a case's elapsed time is spent waiting between operations. If that is so, redesigning the process with fewer operations and shorter queues between them may very well bring spectacular benefits.

But suppose you have an untried new design for a process to assess; if it has less queueing between operations than the old, then an improvement in cycle-time is very likely, but what improvement: 53 days down to 20, to 12, to three? Whatever queueing is contained in the new process may be a substantial part of its cycle-time, but how much will that be? It is relatively easy to estimate the time taken for actual operations, but by no means obvious how to allow for the waiting-times in queues.

Does this matter? After all, if a certain new design is manifestly simpler and neater than the old, it is safe to adopt it even without any quantified forecast of the improvement. Yes, but you don't want just *any* new design that is better than the old; you want the one that is better than all the other plausible new designs. To compare the possible design options it may be essential to make estimates of cycle-times, including time spent in queues.

The most natural way of attacking the problem is to set up a simulation: use software to simulate the passage of (say) 1000 imaginary cases through the process under design option A and to generate statistics about the resulting cycle-times; do the same for designs B, C and D; decide the best design option on that evidence. This kind of simulation has long been used for the study of physical processes, in factories, container ports, pipelines etc — as opposed to the office processes characteristic of re-engineering. The drawback is that simulation is a fairly esoteric field, and many of its experts are quite distant in interests and temperament from the people who normally design or consult about administrative systems.

Thus it may be attractive or even essential to use simulation; but that means entering a strange, rather unwelcoming new field. This is one of the awkward dilemmas that enthusiastic, can-do books on re-engineering tend to evade. The choice is not two-way. The real issue is *how far to go* beyond pure descriptive modelling into quantitative analysis. Certain processes may benefit from quite an advanced simulation study; with others this may not be a good investment of time and skill. It is quite a challenging task to decide on the best-buy approach.

Prototyping

The notion of setting up a crude prototype version of a new design for a process in order to try it out, collect feedback, modify it, try it out again, and so on is attractive. The snag is that a design team can be seduced into a flabby, ill-controlled project. At least, with the traditional, methodical approach of deciding what is required from the system first on paper, and only then developing any software for it, you can make detailed project plans and monitor progress. With prototyping it is harder to be sure where you are, and there is a danger of endlessly modifying a process that is never quite ideal.

Since prototyping is an inherently loose and iterative activity, there is no complete answer to this control problem. Nevertheless, one main reason for ill-justified prototyping overruns is confusion about exactly what the prototype work is meant to achieve: testing for flaws in a proposed design, filling in minor detail, making people feel involved, examining technical feasibility? Some of these motivations could be in conflict. If a clear decision is made about the purpose of prototyping within the design approach as a whole, there is a much better chance of well-directed work that is a good buy in the circumstances of the project.

REPRESENTATIVE IDEAS

This section offers a brief taste of process simulation, and also of the simpler technique of straightforward calculation.

Calculation and Simulation: Representative Approach

Cross et al cover straightforward *calculation* of process performance in a general style that is representative of many consultancies' approach:

Calculation by Spreadsheet

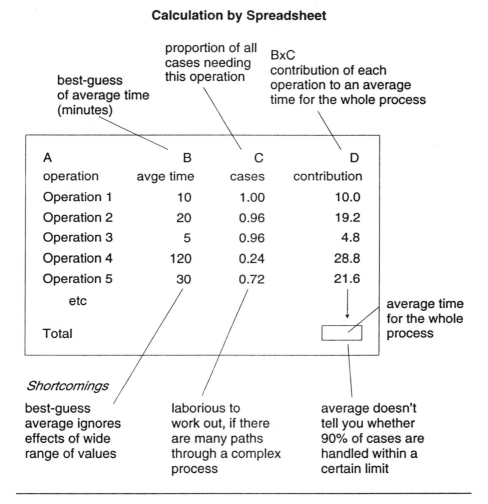

proportion of all cases needing this operation

BxC contribution of each operation to an average time for the whole process

best-guess of average time (minutes)

A operation	B avge time	C cases	D contribution
Operation 1	10	1.00	10.0
Operation 2	20	0.96	19.2
Operation 3	5	0.96	4.8
Operation 4	120	0.24	28.8
Operation 5	30	0.72	21.6
etc			
Total			

average time for the whole process

Shortcomings

best-guess average ignores effects of wide range of values

laborious to work out, if there are many paths through a complex process

average doesn't tell you whether 90% of cases are handled within a certain limit

● Attach best-guess estimates of average timings to each of the (say) 80 operations of the current process. Bring in other data about the workflow; eg, on average, 4% of new cases contain errors; and there are two main types of case, split in the ratio 1:3 etc. Using a spreadsheet, as shown in the diagram, calculate the average time taken by a case to pass through the process.

● Proceed in exactly the same way with the new streamlined process that has (say) only 40 operations.

● Compare the two sets of figures to show how much better the new process is. For example, average cycle-time is currently eight days; with the new process it will be three.

These authors give an example of this approach applied to a company processing five types of orders. The striking results of the calculations are said to have energised company managment about the urgency to change, and to have 'closed the sale' of the re-engineered process.

Cross et al also compare this calculation approach with more sophisticated *simulation*, where software follows each of (say) 1000 imaginary cases through the process, in order to produce overall statistics, such as '62% of cases were handled in four days or less.' Here are their main conclusions:

● Simulation — as opposed to calculation based on best-guess estimates of average operation times and other variables — is overkill in most instances.

● Calculation without simulation is sufficient when there are only a few main paths through the process for a case, with minimal exception routing; and where process and business rules are simple; and where there is minimal variation from best-guess times for each operation; and where work is evenly paced through the day.

● Simulation, on the other hand, is needed where the above conditions don't apply, or where other forms of complexity exist, eg where the processing of different parts of one case on parallel paths has to be synchronised, or there are delays, waits and queues.

● In many service businesses (eg in a bank branch), work arrives erratically and process times vary greatly; with such processes simulation is called for.

The last three of these four points seem very sensible. But if they are, then many typical processes for re-engineering will surely need simulation, and if that is so, the first point must be false.

DISCUSSION

This section concentrates on showing how the varieties of calculation and simulation techniques form gradations across a continuum.

Calculation Approaches

To start with, regard the straightforward calculation approach described and illustrated above as *gradation 1*. The data for these calculations may be laborious to work with in a spreadsheet, if there are many points in the process where a case may branch off in several directions. The example cited by Cross et al, with five types of orders, must have been tricky to get right. But that example raises a more serious question: Did the enthusiastic managers and consultants realise that the logic of their approach rested on some large simplifications?

Cross et al are almost the only people to suggest that it can be inadequate to work with best-guess estimates of average figures, but they merely hint at the possibility. A more sophisticated approach — *gradation 2* — is to use estimates of the *range* of possible values, rather than one average value, for at least a few critical items in the calculation.

Perhaps one particular operation done for a great many cases takes, on average, three hours, but this figure hides a great variety: 95% of all cases take between one hour and five hours; the remaining 5% even more or even less. Why should there be such a range? For an operation such as keying in data taken from a letter, then the range probably will not be that great, and an average figure may well be an acceptable approximation. But suppose the data for a complicated case has to be extracted in a face-to-face interview; it could easily take five times longer to interview an awkward individual with a complex case than an articulate individual with a more straightforward case.

Fitting such subtleties into a spreadsheet soon becomes onerous and may require some knowledge of statistical theory (normal or other distributions, standard deviations etc), but, if done, it can produce findings such as that the average cycle-time will be three days, and 90% of all cases will take seven days or less. This can be important. If the seven-days figure were unacceptable, you might alter the design of the process in such a way that the 90% figure was brought down to six days, even at the expense of an average cycle-time of four days.

A simpler but less useful variant of this gradation 2 is to provide several figures for a critical operation, such as middle (three hours), low (two hours) and high (four hours). If this is done for each of (say) four operations, a spreadsheet can calculate the cycle-time for each of the 81 (3x3x3x3) possible combinations. However, it can sometimes be hard to know just what conclusions can reasonably be drawn from the resulting figures.

Simple Simulation

At gradation 2, three or four operations out of 50 might be thought critical enough to warrant the additional subtlety of defining ranges, while best-guess averages were adequate for the others. But if any more than a few operations need ranges, it becomes easier to go to *gradation 3*, a relatively simple form of simulation.

Here you define for each operation in the process (or for however many of them you like) both an average and some indication of the range (eg operation 1 might be: average 30 minutes, 95% of all cases within the range 20-40 minutes). The simulation software runs roughly as follows:

● Take the hypothetical case 1. How long will it be on operation 1? The system generates a random number: 22 or 31 or 28 etc — but does so in such a way that, if generated for many cases, the random numbers will correspond to the range defined; eg 95% will be within the range 20-40 minutes, and 5% outside.

● A similar procedure is followed for case 1, operation 2, operation 3 and so on. Any branches within the process (eg error paths or paths for different types of case) are also taken in a similar random but weighted way. In this way a total cycle-time time for case 1 is calculated.

- The same procedure is followed for hypothetical case 2, and for (say) 998 more.
- The results of individual cases are not material, but the summary figures for all 1000 cases are. Average cycle-time, or the 90% figure or the percentage of cases handled within 10 days or any statistic in these forms can easily be produced.

This gradation-3 approach may well be easier to set up and yield richer results than gradation 2. Simulation mavens call this and the following gradation (which also uses random numbers) *stochastic simulation*; the approaches of gradations 1 and 2, by contrast, are *deterministic simulation* — really just a pretentious term for arithmetic.

Discrete-event Simulation

But for many processes all these three gradations may still be inadequate. The main snag is queueing time. In the example given by Cross et al of a process handling five types of orders, there must surely be some queues of cases between some of the operations. Unless credible estimates were made for these in the calculations for the new process, the management should not have been energised about the urgency to change at all.

It may be possible to make a reasonable estimate of the time, or range of times, for an operation itself (eg filling in a form), but how can you estimate the time that a case will wait in a queue between one operation and another? Cross et al don't say.

Gradation-3 simulation doesn't touch this problem, since it assumes that estimates can be made of the range of times around an average, not only for operations but for queues too. It simulates cases one by one, without attending to the fact that in reality one case may have to wait behind others in a queue and thus be delayed, perhaps substantially. The time a case spends in a queue waiting to be served will be strongly affected by how many other cases are already waiting in the queue; but how is that to be estimated? The queue discipline may be complicated: suppose some cases, under certain circumstances, can jump ahead of others; how can that factor be brought in?

Before coming to the most attractive response to the challenge of queues in processes, an aside is needed. There is a branch of operations research called queueing theory. An expert in this field will know certain formulae to apply to the study of queues, and — more important — will be able to judge which formulae suit which situations, and to suggest simplified but safe approximations to bypass laborious qualifications that pure theory might require. If, and only if, such a person is available, it can make sense to apply queueing theory to estimate certain queue-times within the process — in combination with any of the three gradations given so far. Even then, this is only manageable if there are just a few critical queues in the process.

Discrete-event simulation, *gradation 4*, is often a more attractive approach. As the illustration shows, you estimate the average time (and their ranges) for each operation, but, rather than estimate queueing times, you define each queue's logic (often but not always, first in, first out). The simulation software then takes a body of imaginary new cases arriving at random times, though at some defined overall rate, and follows their passage through the process. Thus, it looks at the situation (say) minute by minute, to see where each case is: being processed within a certain operation; or else waiting in a queue behind other cases. Then it determines what will happen to each case a minute later (eg complete an operation, move from queue to operation, move up the queue), and so on.

Statistics from the simulation can be very informative: eg average time for all cases was eight days, but 3% of cases took 20 days or longer; or, though it seemed a good idea to give a minority of tricky cases a lower priority, the result is that they average 25 days, and that is quite unacceptable. Moreover, by changing a few details, you can simulate non-typical conditions too; eg if 20% of staff are away with flu, or if it is the Christmas period.

The beauty of this approach is that, given suitable simulation software, you can set up a realistic simulation without possessing any great knowledge of queueing theory.

Discrete-event Simulation: Outline Concept

1. Make descriptive model
of process

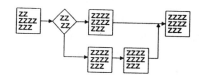

2. Add quantitative and
queueing information

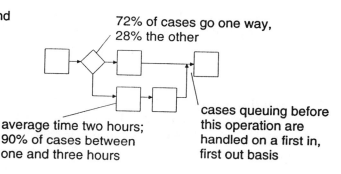

72% of cases go one way,
28% the other

average time two hours;
90% of cases between
one and three hours

cases queuing before
this operation are
handled on a first in,
first out basis

3. With software simulate the
passage of cases (A, B, C etc)
through the process.
At any snapshot moment, some are
being handled by an operation, and
some are waiting in a queue before
an operation.

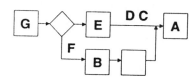

4. Afterwards, analyse results; eg 97% of cases took less than five days.

Continuous Simulation

The description so far is of so-called *discrete-event simulation*.
Software follows (say) 1000 imaginary cases through the process,
sees what happens to each, and summarises the results. *Continuous simulation* is a different technique.

Take, for a moment, a model of the waxing and waning of a
population of fruit flies in an orchard. This might simulate the

lives of thousands of imaginary flies in a discrete-event simulation, but there is a much neater approach.

One definite factor affecting the population level today is the level of the population yesterday. (Other factors may be temperature, the amount of food available etc.) Thus, to be slightly more abstract, a formula bringing together all the variables affecting the population level at time *t* will include the population level at *t-1* or some such earlier time. Start off with a given population level on (say) May 1; feed that figure as one variable into the calculations for May 2; having worked out the figures for May 2, proceed to the next day; and so on through the simulated summer.

In a similar way, as the diagram suggests, within a local government planning process the formula for the number of building-permits waiting to begin a certain operation will contain, among other variables, the number that were already waiting the day before, the number that completed the previous operation of the process the day before, the number that failed at a later stage of the process and have just been fed back in revised form, and so on. Work out this equation for one moment of time, and that result can be fed into the same calculation (and to other calculations about other parts of the process) with respect to the next moment, and so on. Do this repeatedly for closely spaced times, eg every day over several months, and the result is a kind of continuous simulation of the flow of cases through the process.

Unlike gradations 3 and 4, continuous simulation doesn't look at individual imaginary cases. It adopts a more holistic approach, and so might be called (though rather loosely) *gradation 5*. It is suitable for processes where the relations between operations are relatively small in number but complex. For processes where these two conditions don't apply, such as the process for a new insurance policy with 50 or 100 operations, it is less suitable. For them a continuous graphic view of peaks and troughs over time is much less interesting than overall statistics about average cycle-time, or the 90% figure, or the percentage of cases handled within a certain time; also, while the model has many operations and flows, few of them need to be defined by intricate mathematical formulae. Thus, for most re-engineering projects, gradation 5,

Continuous Simulation

Suppose a certain process is rich in feedback and complex interaction:

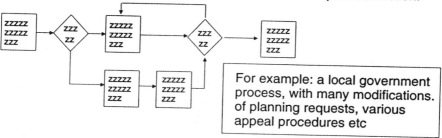

For example: a local government process, with many modifications. of planning requests, various appeal procedures etc

Then it may be possible to define formulae for key elements in the process:

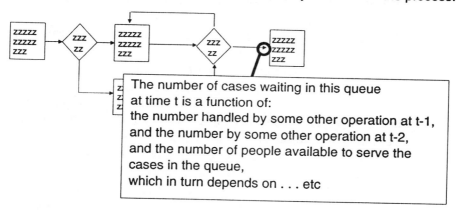

The number of cases waiting in this queue at time t is a function of:
the number handled by some other operation at t-1,
and the number by some other operation at t-2,
and the number of people available to serve the cases in the queue,
which in turn depends on . . . etc

These equations can be solved for successive moments of time over (say) an imaginary four-month period, to give output like this:

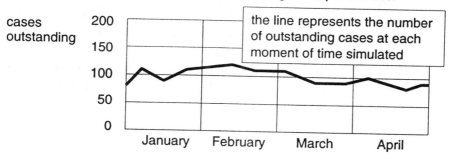

cases outstanding

the line represents the number of outstanding cases at each moment of time simulated continuous simulation is less interesting than gradation 4, discrete-event simulation.

Kinds of Process, Decisions about Simulation

Analysis by kind of process can assist decision-making about the use of simulation:

- With some **transaction-based processes**, the queueing features needing simulation may be present but not crucial. However, simulation is particularly apt for a transaction-based process with high volumes of cases and complex queues; ie a case waits in a queue and goes to whichever out of (say) five people (servers in the jargon of queueing theory) is free first. Here, slight differences of case volumes or numbers of servers can make large differences to queueing times, and usually the way to study this is by gradation-4 simulation.

- **Matter-based processes** often need simulation; their unit of process is larger, more complex and more variable, and so awkward queues can arise. For example, a building-permit process may include features such as the architectural or legal expert only available every two weeks, or the construction engineers who, once started on one case pursue it as far as they can before tackling the next. Thus, if matter B arrives in a certain queue five minutes after matter A, it may wait for five days for attention. (You may think that the mark of a foolish process, but to prove that contention you ought to demonstrate some other design that, on balance, would be better. To do that credibly you might have to run a simulation.)

- By the nature of **project-based and facility-based processes** simulation scarcely arises as a possibility. You can't really make a model of a process that stores engineering drawings to facilitate product development at Kodak, simulate its performance on development of 1000 different camera products, and summarise the results statistically.

Technology for Calculation and Simulation

Many software products intended for process modelling can store quantitative data and use it for *gradation-1* calculation. This avoids the need to keep a process model and a separate spread-sheet consistent. However, hardly any of the process-modelling software products directed at the re-engineering market handle the gradation-2 or higher possibilities summarised in this brief-ing.

The main way to achieve *gradation-2* calculation, ie being more sophisticated than working with averages but without going as far as simulation, is to use a spreadsheet, taking advantage of its macros and special functions for statistical calculations. Some care and knowledge is needed to use them without committing solecisms of statistical theory. For *gradation 3*, the spreadsheet needs to be reinforced by special add-on software.

For *gradation 4*, discrete-event simulation, there are three main possibilities:

● Write software in a **programming language**. It would be absurdly laborious to use a well-known language, such as Basic or Cobol, but a number of specialised languages exist, that are rich in powerful instructions for constructing and running simulations, eg SIMSCRIPT, SIMULA, CSMP and GPSS.

● Use a general-purpose, **graphics-based software package**. Instead of writing program code you draw a process model on the screen, and then make choices from menus to define all the detail affecting simulation. Thus, for each operation of the process, you access a menu to give the best-guess time and define how that value can vary: the menu may allow you to choose between a uniform distribution over a certain range, normal distribution and other variants. This kind of package is general-purpose: it can be used to simulate an insurance company or a prison or a container port.

● Use an **industry-specific or process-specific simulation model** — probably with some modifications or at least adjust-ments of values to suit your own organisation. For example, you may be able to use the same simulation model as some other insurance company for the new-policy process; or a consultancy specialising in insurance may offer a general new-policy simulation model with built-in options to suit various ways of handling the process; or a software supplier may offer a generic simulation model for *any* new-contracts process: not just an insurance policy, but perhaps for equipment maintenance or leasing or other cognate types of contract. Such a ready-made simulation model may have been built in either of the two ways above.

Using a graphics-based package has obvious attractions over writing program code; the latter is only a serious option in the unlikely case that the simulation needs many complexities beyond the scope of any general-purpose package. The third option, a process-specific simulation model, is alluring — if available. But it probably won't be, since simulation of administrative processes is a rather new field. However, that may change over the next few years. For the moment, at least, the second option, the general-purpose graphics-based package, is much the most germane.

There are quite a few candidate packages, but most or all were developed for simulating industrial rather than office processes. Some software suppliers have not yet realised that their products may be used in re-engineering projects too. This matters because you generally need advice from the supplier's specialist staff to get the best out of a sophisticated product. Three products whose suppliers have already recognised the re-engineering market and pursued it actively are WITNESS, SES/Workbench and Taylor II.

For *gradation 5*, continuous simulation, the same three options exist as for gradation 4. As argued above, continuous simulation is less applicable to most re-engineering projects than discrete-event simulation. However, one general-purpose package strongly marketed for continuous simulation in re-engineering is ithink (sic).

Gradation-4 Simulation Software:
Example Features

Using a general-purpose, graphics-based software package for gradation-4 simulation is an attractive approach in many re-engineering projects, but, at least at present, a moderately-to-highly ambitious one. There are some more issues to tackle, but first, it is worth seeing a lttle more of the characteristic approach.

As the diagram shows, the essential principle is to draw a graphic process model on the screen, and then, through menus of standard choices, add in all the hard data needed for a simulation. Thus:

● For each operation-box in the model, you give **operation-timings** data, such as a best-guess of the time needed and definition of the range of possible values — uniform distribution, normal distribution, Poisson distribution and perhaps a dozen other esoteric variants may be offered by the package.

● For each operation-box in the model, you define the **queueing discipline**; ie the logic which determines which case, if there are several waiting, should be handled by the operation first. Here is an impression of the range of possibilities: wait until server is free, cases are processed in strict sequence; *or* wait until server is free, but not processed in strict sequence (eg last-in-first-out, or round-robin, or random etc); *or* strict-sequence, except different priorities for different types of case; *or* wait till a certain time, eg end of day, or next monthly review board meeting; *or* wait till some non-temporal condition is met, eg until a reply to a letter for clarification is received. Various combinations of these are also imaginable.

● For each queue before an operation, you can define the **server discipline**: eg any of three people are available to carry out the operation, or, more complex, the number of people available for the operation varies according to the time of day.

● The model will contain junction-boxes to control the routing of a case through different branches of the process. Again the **routing logic** can be defined through a menu: eg a case branches one way or the other according to its type; or at random but

Simulation Package: Graphics-based, Menu-driven

To select process elements and define their details by menu is vastly easier than the alternative of writing code in a simulation programming language.

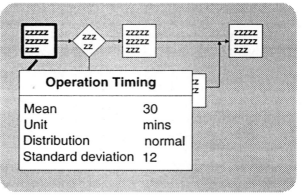

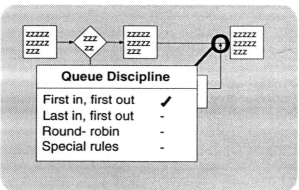

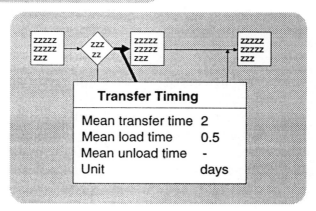

according to some overall probability; or according to some simple condition (eg time of day) etc.

- Figures can be attached to the **transfer timings** between operations (eg for a letter in transit in the post), with, if necessary, loading and unloading timings (eg time spent in a postroom at each end).

- Through another menu you define the **volume of input** to be simulated; eg simulate 5000 cases, split over three types of case in a certain ratio, arriving at random intervals throughout one week, but with a certain pattern of peaks and troughs.

- Through another menu you define the **output analysis** required from the simulation: average cycle-time, number of cases within a certain limit, maximum figure for 95% of cases, etc.

In principle, a good graphics-based simulation package allows you to set up and run a powerful simulation, by drawing diagrams on the screen and making choices from menus, without having to write any program code. It also makes it easy to run a number of simulations with slightly different variables.

Gradation-4 Simulation: Issues and Tradeoffs

The first issue is *non-standard logic*. There may be features needed in the simulation that can't be defined through menus. Suppose, as a wild example, the discipline of a certain queue is affected by whether or not it is the server's birthday; you can scarcely expect a package's standard menu for queue discipline to include a 'server-birthday' option. Since there is always a chance that some such non-standard piece of logic will be required, a package provides a facility for defining any non-standard logic — either in (say) the C language or in some package-specific format.

In many processes there is little or no need for such coding. But the topic can loom large for the following special reason. The simulation model is defined by logic at the level of each of its individual parts: for each of the individual operations, queues, servers, junction-boxes and so on in the model, menu choices are selected or, exceptionally, code is attached. But suppose logic is required that implicates *several parts* of the process: eg the number of servers available to handle operation 4 is not fixed; it depends on how busy the same people are in handling operation 28, which in turn depends on something else ... This kind of thing may be quite tricky to fit in, and thus leave a delicate choice between embarking on a demanding and thus risky project and accepting simplifying assumptions into the model.

The example just given is related to a second issue, *people-specific features*. Simulation of an industrial process is concerned largely with finding the most cost-effective configuration of machines and other objects; eg optimising the number and arrangement of cranes, storage areas and guided-vehicles in a container-port. In an administrative process, clerks rather than cranes and customers rather than containers are being simulated. This brings two main complications:

● People are far more **flexible** than machines. A crane is either being used or is idle, but if there are no new claims to process, the insurance clerk can switch to lower-priority or less-essential tasks: filing, studying, making extra personal phonecalls to increase (it is hoped) customer satisfaction etc. This ought, ideally, to be represented in the model, but can easily get very complex. Again the choice between being ambitious and accepting simplifications arises.

● People have **feelings**. For a container-port the optimum design solution is (more or less) the one that provides the best balance overall between cost and speed of handling. The facts that a few containers may have to wait in a queue a very long time, or that one crane may have to work much harder than its colleague are irrelevant, because objects have no feelings. But suppose that the optimum new design, on balance, for some administrative process meant that often a customer had to spend 30 minutes in a waiting-room before a certain operation; suppose that with an alternative design there were waits of 15 minutes each before two operations. Would not most customers feel, perhaps irrationally, happier with the second variant? If so, isn't that variant a better choice, even if, on pure cost or throughput grounds, it is slightly less efficient? But once this kind of subtlety is brought in, the definition of the problem and choice of variant solutions to simulate becomes more complicated.

A third issue is *animation*. To see some demonstrations of software products you would think that the main purpose of simulation was to show an animated, symbolic representation of the process — with little insurance-claim icons speeding along arrows and waiting in queues to be dealt with by insurance-clerk icons, and so on. But animation doesn't play a large role in promoting rational decision-making. The summarised throughput figures from the simulation run are what really count. They, not the animation, are the main help in deciding between one option for process design and another.

Nevertheless, animation does have benefits. It is an efficient way of seeing quickly whether the process design contains any gross blunders. It may also stimulate new ideas for improvements. Above all, it may bring soft, change-management benefits, by impressing people outside the main project-team, and making them feel enthusiastic, committed etc.

There is a tiresome fourth issue: *pure process modelling*. Suppose you have used one of the many software products described in the briefing on process modelling: BPwin or RADitor or Action Workflow Analyst, for example. You then decide that you want a simulation, using WITNESS, SES/Workbench or Taylor II. Since all the latter have their own modelling conventions, you will probably have to set up a complete new model in the form required — a laborious and error-prone task.

This can, of course, be avoided if all the modelling is done from the very start with the simulation package. But different modelling conventions expose different facets of a process; you may be particular keen on the insights provided by (to give a random example) an Action Workflow Analyst model, and reluctant to forgo them. One possible compromise is to draw a detailed model (100 operations, say) in the process-modelling package of your choice, and then translate it into a much tighter form (20 operations, say) for simulation, exposing the most sensitive parts (bottlenecks, operations with widely varying times etc), while summarising the rest.

REPRESENTATIVE IDEAS

Prototyping is a long-established technique in system development, with considerable relevance to re-engineering. Surprisingly few authorities on re-engineering say much about it.

A Rare Mention of Prototyping

Davenport's book is clear enough about what a prototype is, as opposed to a pilot, but largely misses the chance to present a stimulating account of prototyping issues. There is a list of eight types of supporting software technology, but this has little or no decision-making force, since there is no account of how some are more suitable than others under various circumstances.

Davenport does include one excellent point, that is ignored by other authors. Here it is, expressed more fully:

● Using software tools to prototype a new process (ie experiment with a partial version of the process and refine it) is one thing, and using software tools to develop the real process that will be operational for years to come is something else.

● Some software tools, eg certain fourth-generation languages, may be usable for both these two purposes. There can then be a choice: use the same technology for both purposes, or use different technology for each.

● Using the same software technology for both purposes will make it easier to move on from the prototype to the operational version of the new process. Even if the transition is not entirely smooth, it is likely to be quicker than if the other option is taken. Moreover, the option of using two software tools for the two purposes has a rather undesirable psychological (ie change-management) factor: you seem to throw away the prototype and start software development all over again.

● But on the other hand, the best software tool for a particular prototyping task may not necessarily be the best for developing the operational system. After all, different purposes may call for tools with different qualities. One that makes it very easy to generate attractive screen designs during prototyping may not have all the features to optimise the response time of an operational system handling high volumes of cases. It certainly can't be assumed, as a general maxim, that using the same software tool for both purposes will always be the best policy.

● In fact, more likely than not, you will face a tradeoff decision: *either* use the same tool for both purposes, even though it is not the best for both, but gain a relatively swift transition from prototype to operational system, *or else* use two different tools, the best you can find for each purpose, but accept that, once prototyping is over, starting up development of the operational system will take longer.

Petrozzo and Stepper merely recommend using the same technology for prototype and operational system, because of the advantages summarised above, and they completely ignore the opposing considerations that properly belong in the calculus of tradeoffs.

This briefing doesn't go into the software technology for prototyping, since that is unavoidably part of the huge topic of software development in general. A useful reference as starting-point for study of the field is: Linthicum, DS. 'RAD Tools: RADical Development', *PC Magazine*, November 8, 1994, pp. 153-210. This mentions, with varying degrees of detail, about 35 software products.

DISCUSSION

As with so many issues the productive advice about prototyping is that there are certain generic options, and that different situations call for different choices.

Prototyping: an Example

To show the issues at stake an example is useful. Take a draft redesign for a magazine's subscriptions process — a vast improvement on the old, except, it emerges, if a subscriber writes in with two queries in the same letter: 'I did not receive the June issue, and you sent me a renewal even though I renewed my subscription last year for two years.' Suppose the new design for the process doesn't allow for this possibility.

Once discovered, the simplest solution may be to add an extra operation: somebody checks every letter for double content. If necessary, copies are made and sent along different paths, just as if two letters had been received. But then two separate replies will be sent out, and that could easily puzzle the customer. The obvious but awkward remedy then is to have some recombination operation later on in the process. This probably requires a sorting operation near the start: ordinary letters are separated out from multi-point letters, which in turn are copied and split . . .

But there can be letters where it is uncertain whether the points made are separate or connected: 'I did not receive the June issue, and I have noticed that you always get my postcode wrong, although perhaps that is immaterial, since the magazine has always arrived safely before.' Now the idea comes up: why not bypass all such problems by storing the customer's letter as a document-image accessible through any terminal? That will make it far easier to devise a neat process — though it will mean quite an investment in technology.

But before that can be decided some account must be taken of complications on another front. The new idea of sending personalised answers to queries (ie automatically printing a clearly written signature) has its drawbacks, since customers may address letters to the person not the magazine (perhaps a problem if one person deals with several magazines, or if responsibilities are altered), or the customer may misread the person's signature. Would document-imaging make things better or worse? Not all customers put the addressee at the top of their letter. Perhaps the envelope should be scanned as an image too . . .

The point is that, however shrewd and bright the people in the team, this kind of iterative, discovery process is inevitable. It is not satisfactory for large discoveries to occur at the pilot stage when the process is already handling live cases. The psychological effects will be very bad, and besides, the freedom to change the process is too limited. It is scarcely feasible at that point to redesign everything all over again to take advantage of imaging technology. The aim must be to make all the crucial discoveries and most of the lesser ones during the design work, rather than later. Prototyping is one of the main techniques for achieving this.

Prototyping: Approach Options

There are at least four generic options:
- *either* **don't use** prototyping;
- *or* get a long way towards firming up the main features of the new process; then, not expecting these features to be overturned, use prototyping to study and iterate through details of **third and lower orders**: roughly anything that can be changed in one corner of the process without having implications for consequential changes in other corners;
- *or* firm up the main principles and general outline of the new process; then, not expecting these main principles to be overturned, use prototyping to study and iterate through details of **second and lower orders**: ie almost everything about the process, apart from such basic principles as whether or not to store letters in document-image form;
- *or* use prototyping almost from the start at all kinds of design levels, including exploration of **first-order** principles; eg experiment with document-imaging before deciding whether to adopt it or not.

Prototyping: Motivations

This analysis is only a start; things are more complicated. Choice between the options should also allow for the fact that there can be several possible motivations for prototyping:
- Generating **bright ideas** for improvement of the design, and also making small enhancements (eg arranging the items of data displayed on a screen in a way that is easier to read).

● Rooting out **objective flaws** in the design. With the new design for magazine subscriptions, there may be some possible combinations of events (perhaps involving letters that cross in the post), that produce wrong or otherwise unacceptable consequences.

● Gaining insight into **sociotechnical factors** about the design. Davenport gives the example of the American tax authority that introduced a process, without objective flaws and in theory much more efficient. However, many staff hated their new jobs (far less personal contact, far more work at phone and terminal) and resigned. The organisation would have done far better to prototype the proposed working-methods first, to study people's reactions to their new tasks. This isn't just a matter of checking for negative reactions; some features may turn out surprisingly popular.

● Encouraging wider feelings of **participation** in the design effort. The three motivations above are about finding a good process design. Prototyping may also be a change-management resource. If staff outside the design team are allowed to participate, they may (depending on the circumstances) develop positive feelings towards the whole change initiative. In some circumstances, it could be wise to conduct extensive prototyping for this reason alone, ie even though it was not considered a good way of generating ideas, discovering flaws or studying sociotechnical factors.

● Examining **technical feasibility**. Part of the design may depend on innovative technology. Perhaps it is vital to know whether the advanced handwriting-recognition technology employed will achieve accuracy rates of 99%, 95% or some other figure. Prototyping may be the best way to find out.

As usual with this kind of breakdown, the essential thing is to clarify motivations, and organise prototyping accordingly. For example, if the participation factor is predominant, then it may be best to allow many people with varying abilities to try out the process. If, on the other hand, rooting out objective flaws is the paramount aim, then the opposite (a smaller number of exceptionally shrewd people using the prototype intensively) may be better. If you don't clarify the motivation and consider the tradeoffs, the results could be disappointing. Inviting dozens of sceptical people in order to build up their feelings of participation, and, in the same prototype, experimenting with all kinds of technical feasibility parameters is unlikely to work well.

▼ Writers who gush about the importance of human factors often muddle together two distinct things. Sociotechnical design is concerned with the human factors of *the new process itself*. Change-management factors are human factors concerned with the *introduction of the new process*. Thus, a process where people spend all their time at a telephone in a cubby-hole may hit sociotechnical design problems. Imposing a radically new process (even with beautiful sociotechnical features) by fiat without any consultation raises change-management issues.

This distinction is useful in much writing about management beyond re-engineering. An article in *The Economist* in January 1993 reported on a Toyota factory in California where new systems based on strong segmentation and standardisation of tasks (low marks for sociotechnical design) were designed by the factory workers themselves (high marks for change management), rather than by white-coated engineers, stop-watch and clip-board in hand. The results, by all accounts, were good. ▲

Kinds of Process, Decisions about Prototyping

Here are some considerations that may affect policy on prototyping in any particular situation:

● If the re-engineering project deals with a **transaction-based process**, you can, in principle, describe a possible new design and demonstrate in a rational way that it is better than some other option, as Cross et al advocate, by making calculations or simulations based on quantitative factors. You can go a fair way towards separating good from less good design ideas in this way, before setting up prototypes.

● With the typical **project-based process** for development of a new product the great design challenge is rigorous integration and co-ordination of the content of engineering drawings or such material. This is practically impossible to prototype quickly and realistically.

● But **matter-based** and some **project-based processes** are often largely concerned with helping people at several locations to interact better in less structured ways. For these, prototyping may be essential, and far more apt than calculation or simulation. Suppose electronic communications were provided between Verdi the composer and Boito the librettist. Would they get their operas ready quicker? Or produce higher quality operas? Who can say? They might do worse; increased communication could lead to more arguments, more discarded drafts and so on. It is very difficult to know beforehand that any one design option will necessarily be better than another. The main way of finding a good design here is by prototyping.

● Processes that offer the user considerable flexibility (many **matter-based** and **project-based**, and practically all **facility-based processes**) pose another difficulty. Experience of sophisticated cameras, VCRs, or phone systems with extra facilities built in (not to mention flexible manufacturing systems) shows that more options can mean a worse system. It may be more demanding to learn and use, and the majority of people may stick to a small subset of the options, anyway. Here the design challenge may be to find the right balance of simplicity and flexibility. Prototyping may well be the best method of solving it.

CONNECTIONS

Briefing 4 explains the difference between pilots and prototypes.

This briefing is closely connected to Briefing 6, which discusses varieties of process modelling and related issues. One great benefit of testing out a new process design by simulation or by prototyping is that variants of the design can be readily conceived and examined; then the process model has to be amended accordingly. Thus the activities described in this briefing will often be mingled with those of Briefing 6.

8.   IT Impact

Some books and some consultants make innocuous references to the role of information technology (IT) in re-engineering, without suggesting that it can call for tricky judgements and far-reaching decisions. This book argues that, though situations differ, IT ought *frequently* to be a complicating factor in re-engineering design work. From this it follows that on *any* particular project, you should make an early judgement about the attention to be given to IT, and take decisions of design approach accordingly. On a certain project you may give IT factors a minor role, but this should be a conscious decision based on explicit, situation-specific arguments. A good start towards explaining and justifying this position is to divide the theme of the impact of IT into three separate complexes of issues.

Technology Stress

The controversy about the place of IT in re-engineering is best illustrated by two opposing opinions:

● The primary task is to design a new process in business-oriented, non-technical terms. Software and hardware may be needed to support the new process design, but working out that technology is essentially a question of filling in the detail of process features already settled.

● No, you should not make a non-technical design first and fill in technical detail later. Many of the best design ideas to be found are based on deft application of up-to-date technology. Therefore exploration of technology possibilities should be encouraged as a stimulus to innovative process design.

The second position may seem likely to cover the range of design options more thoroughly and lead to more innovative ideas — but perhaps only at the considerable price of making the whole design task more challenging to carry out and more arduous to co-ordinate.

Why does this debate matter? Most obviously it will influence the makeup of a design team. There is a strong influence too on the structure and rhythm of design work as a whole. If IT is seen as a way of supplying whatever non-technical demands are established, it is natural to work on the demands first before considering the supply; then the project may take on a pronounced top-down, step-by-step character. But if a more complex interaction between supply and demand possibilities is recognised and encouraged, then a more subtle, iterative structuring of the design activities is needed.

Significant Technologies

When IT considerations are allowed a forceful role in re-engineering, another interesting area for debate emerges: Are some particular technologies more applicable to re-engineering than others? If so, which technologies, why, and what implications does this have for decisions about the approach to the re-engineering design work?

IT Constraints

There is a third complex of issue: IT constraints, and most formidably, technology infrastructure. Many organisations have a database or telecoms infrastructure that transcends individual application systems. The database and network are painstakingly designed at the outset to serve many systems in many departments, and, albeit to a difficult-to-define degree, to have the flexibility to accommodate new needs in the future.

If any large-scale new system is set up *outside* this infrastructure, the costs in time, money and complexity may be exorbitant. Moreover, even if that price is acceptable, the whole infrastructure concept is undermined, and recovery of the original infrastructure investment is made more difficult.

But unfortunately, if a piece of re-engineering is truly radical, then it is quite likely that the design of the new process won't match very well against the assumptions of those who designed the existing infrastructure. After all, if it does, it probably won't be radical.

Therefore to implement a radical new process you may have to build from scratch outside the organisation's existing IT infrastructure — with all the unpleasant effects just outlined. Or this prospect may seem so appalling that, as a lesser evil, you accept the constraints of the infrastructure and make the design less radical and achieve lesser benefits. That is the quandary, painted with a broad brush. It is certainly important enough for careful exploration.

REPRESENTATIVE IDEAS

On technology stress there are in fact more than just the two positions given above to be found. Moreover, one sign that this is a tricky issue is that some of the published advice is confused and contradictory.

The Non-technical Position

BAI, a bank described by Hall et al, exemplifies the non-technical attitude as vividly as any. The bank's approach was to design the new process in full non-technical detail first. After that the design was handed over to the technology team, charged with thinking through the IT implications: 'We had the technology team work independently of the organization team so that current system limitations did not influence the organization team's redesign.' Then, while the technology people were developing a client/server architecture, non-technical people were working separately to decide on staff changes, define new skills and job responsibilities, and design a new physical layout for bank branches.

BAI may be a success story, but that can't deflect the general question of whether such compartmentalisation is safe. Can you really design a radical but feasible cheque-deposit process without taking account of the strengths and limitations of technology such as document scanning, imaging and electronic filing, and expert systems? Is it right to design the physical layout for bank branches in the organisation team, while people in the technology team are still deciding whether there should be one large printer or several small ones, or whether expensive, high-resolution VDU screens should be used or not?

Cross et al offer eight pages on IT out of 300 without conveying much indication of where in their 11-step methodology, IT makes an impact on design and decision-making. That they hold an *attitude* rather than a well-argued position is shown by their example of two insurance companies. Company A invested in imaging and voice technologies, without making much change to procedures, while B invested in radical redesign of procedures, without extravagant use of new technologies. Naturally B achieved greater benefits. This is a pointless comparison, since nobody would seriously doubt that A's was a foolish policy. The example would only have any point if there were a company C that both used new technologies and redesigned procedures; the comparison of B and C would then be of interest.

In fact, the simple non-technical position is rarely expressed in a cogent way. Many articles contain statements like: 'IT should remain a tool which supports and enables re-engineering, rather than a goal in itself.' But this is empty as a recommendation since nobody does advocate that IT should be a goal in itself. Worse, it muddles the real debate, about how 'tool', 'support' and 'enable' should be understood in this context.

▼ The false antithesis is a version of the disreputable *straw-man* technique: setting up a distorted version of an opponent's position in order to have something easy to attack. It is a mis-representation to imply that somebody who believes that new technologies may assist re-engineering necessarily intends to apply technology without considering radical redesign of procedures. The strawman lurks in many articles about management and in fields beyond too. ▲

Moderate Technology Stress

Davenport and Short provide a neat clear statement of their position on the role of IT in re-engineering. They say: 'awareness of IT capabilities can — and should — influence process design' and 'IT can actually create new process design options, rather than simply support them.' Within their standard approach one out of the three steps concerned with design is to brainstorm through ideas generated by IT possibilities. As a brief example, they point out that a team redesigning a company's product development process ought to have some awareness of the possibility of transmitting computer-aided designs from one location to another.

In the later interview Davenport argues that to design a process in a non-technical way first, and consider IT afterwards is like designing a new building without knowing what kind of building materials are available, what kind of elevator technology, and so on. In the book Davenport makes the same point with another analogy: 'A sculptor does not take a design very far before considering whether to work in bronze, wood or stone.' But the Harvard style reduces the impact of the point. Rather than use the expanse of a book to explore perplexing tradeoffs for the way redesign should be approached, Davenport takes the tamer course of giving numerous brief examples of innovative applications permitted by IT.

▼ Litanies of examples, whether illustrating IT applications or any other facet of management, make easy reading, but may not help the real-life manager very much. It is more instructive to spell out, with supporting examples, how certain generic factors tend to conflict, and how the challenge is to find the best balance of tradeoffs in a given situation. For example, the desire to make optimal use of available technology possibilities may perhaps conflict with the aim of minimising the project-risk, or with taking a methodical approach, where means and ends are kept clear. A guide to quandaries is far more helpful to the decision-maker than any number of neat success stories to be admired.

Obeng and Crainer also seem to be on the side of technology stress but their statements are of more interest to the student of argumentation analysis, than of re-engineering: 'For IT to play a full role in enabling re-engineering to happen, it has to be involved in the process from the very beginning.' This almost smuggles in the assumption that IT *ought always* to play a full role and be involved from the beginning. To see this, replace IT by needlework or geomancy or any other term whatsoever, and the statement will still be roughly true. The nefarious near-tautology is a common snare in writing about management. ▲

Extreme Technology Stress

There is a more extreme pro-IT position. Hammer and Champy point out that often people do not know they want something until they see that they can have it; then they feel they can't live without it. Thus insight into a certain technology may generate needs and uses not hitherto articulated. This, they claim, is a variant of Say's Law in economics, about supply sometimes creating its own demand. They go on to press the argument to an extreme.

At any given moment, certain technologies are really-new, ie only just available in a viable form, ready for the transition from R&D to serious commercial use. Hammer and Champy urge you to plan ahead to apply a really-new technology in your re-engineering the moment it is feasible. To do this you will have to make detailed plans based on *predictions* of technology advances over the years ahead. In fact, you can even stimulate technology suppliers by nagging them for the new technology you would like. Brief examples of this nagging technique are given: Chrysler with satellite telecoms and American Express with document imaging.

Hammer and Champy's book isn't entirely satisfactory on this technology-stress issue. If the authors really do hold such a strong position on technology stress, why don't they devote far more attention to IT? Why do the eight main examples of re-engineering in the book show rather little adventurous use of new technology?

Above all, why is there nothing about the challenge of structuring the re-engineering project so that Say's Law about supply stimulating demand can operate effectively? There is an awkward conundrum: on the one hand, team-members are to be encouraged to seize on attractive supply factors and search for plausible demands they might meet; on the other, it is desirable to maintain a sense of direction, minimise speculations that prove fruitless, and avoid indulgence in beguiling but not viable technologies.

Variant Views and Some Confusion

The book by Carr et al is in a broad-brush style, but it sketches out another plausible-sounding standpoint:

● It is usually stupid to take an existing process, and apply technology to it, without improving the non-technical procedures as well.

● One way to make a new process is to design its procedures first and work out the IT details afterwards. In fact, much re-engineering actually practised is like this.

● The alternative is to 'create a synergy between process redesign and information technology'.

● It is best to decide at an early stage whether the second or the third of these approaches fits the particular project. Deciding that point explicitly is an important step towards structuring the design work, deciding which people to involve, and so on.

Though this sounds sensible, the book can be criticised for lack of clarity or even muddle. It says that *if* IT is to play a big role, then IT people should be involved throughout the 'Discover' phase; but, the question then arises, since 'Discover' is the first phase, at what earlier point are you supposed to decide whether or not IT is to play a big role?

Johansson et al avoid the issues of this briefing entirely. The book by Morris and Brandon is almost unbelievably incoherent on this subject. The account of the activities in the nine-step re-engineering methodology makes virtually no mention of IT. But a separate chapter does describe various IT-related activities: designing corporate technology architecture, finding information- and technology-critical areas, etc. How that work fits into the step-by-step methodology is not explained. As it stands the methodology is unusable, for its muddle on this issue alone.

▼ If it is true that these authors have it all wrong, why bother to mention them at all? The point is that, though you never read those particular books, you may well come across consultants purveying their own approaches to re-engineering. The issues where the books run into difficulties tend to be the troublesome ones for other consultants' methods too. In any field of management, finding the weak points in the books is a good way of discovering where the inherently tricky issues lurk. ▲

DISCUSSION

This section contrasts the representative ideas in order to develop a line of argument and expose options for decision on the issue of technology stress.

New IT as Driver of Re-engineering

Many writers on re-engineering say that it has recently become possible to make huge improvements in business processes. But huge improvements, if feasible, are welcome at any time. Why should they have suddenly become feasible? Hammer and Champy mention certain megatrends: increased customer power, intensified competition between suppliers, a new state of constant change, and IT. The first three are the kind of impressionistic factors that management gurus always bring up to show that this is an exceptionally challenging time for business, calling for a whole new set of buzzwords — just as any barber will always find reasons to suggest that now is the right time to have a haircut.

But the IT argument is more compelling altogether; there is incontrovertible evidence to support it. In 1980 it was not affordable to put a computer on every employee's desk, but in 1990 it was; that change is undeniable. Davenport and Short are clear about this: 'information technology's promise — and perhaps its ultimate impact — is to be the most powerful tool in the twentieth century for reducing the costs of . . coordination.'

As the previous section shows, many books and articles about re-engineering don't make claims about IT that strongly. But, unless such claims are valid, the bold rhetoric is little but banal exhortation to do away with pointless work-practices, be imaginative rather than dull, look out for ways to make a process simpler, and so on. The point about the power of newly-affordable IT is essential to rescue any author from charges of asserting the obvious. This suggests that the merit of any standpoint or methodology that omits or underestimates the complications brought by IT factors is doubtful.

But wait; is it really true that IT is not merely cheaper but has suddenly reduced the costs of co-ordination? The crucial claim is that around 1990 — rather than five or ten years earlier — technology became particularly suitable for designing radically innovative, because differently co-ordinated, processes. Is this really true? Here is some support for the view:

● Several technologies became both mature and affordable during the late eighties.

● These technologies are much more powerful in combination than singly. This synergy makes radical change possible.

● For example, take three features of the re-engineered Ford procurement process: information is accessed and updated in the database easily and flexibly; information is widely available through a telecoms network; the setup is fault-tolerant, ie all kinds of recovery and control mechanisms are built in, invisible to the user, to allow work to continue even in the face of technical mishaps or other complications. Only with these three technology features together is it feasible to rely on the concept of information (eg what parts are on order) being held electronically once in the database, rather than on several pieces of paper and their copies. Given that, radical redesign becomes possible.

● Thus it is indeed a plausible claim that the information technology central to the Ford example, and many like it, reached a decisive level of maturity and cost-effectiveness in the late eighties.

One other technology factor is less prominent in the Ford example, but fundamental to many others: the arrival of cheap PCs. Once there can be a powerful PC on the desk of every office-worker, all kinds of new possibilities suddenly arise.

Choosing between the Positions

These considerations, together with the fact that the protagonists of technology-free re-engineering never put any convincing line of reasoning together, surely establishes the position of Davenport and Short rather than Cross et al.

But this is not to go all the way with the extreme concept of planning uses for new technology before it is even available. First of all, since nobody can forecast all, or even most, technology developments accurately, much of the effort is certain to be wasted. The activity is rather like that of a country's general staff drawing up plans for military operations against a variety of possible enemies, many of which may never be put into effect. Chrysler and American Express are large corporations. Possibly it is cost-effective for their share-holders to finance this kind of thing, though it would take more information than Hammer and Champy provide to establish that. But for most organisations, even large ones, the challenge of making good use of the full range of already established technologies seems daunting enough.

▼ This seems to be an instance of a certain generic fallacy. Obviously, it would be pleasant to implement a striking new process the moment advanced technology became available, and to carry it off without any hitches. And, if suddenly attacked by a tiger, it would be nice to get your knife in about 3cm below the animal's right shoulder-blade. But to say that a certain objective is desirable, is not to show that it is feasible.

There is a related fallacy: that of confusing aims and advice. Somebody who dismisses a book on the subject of how to be happy is not necessarily against happiness; more likely, the book is judged to contain no useful advice on how to achieve the aim. Similarly, to disagree with a person's views on how to take advantage of IT is not to be against the aim itself — only the advice offered for achieving it. ▲

IT Knowledge Issues: an Example

The views of Davenport and Short are surely sound, but their article offers the merest sketch, and the implications of their position can easily be overlooked. They give the simple example of a manufacturing company re-engineering its product-development process. Unless somebody in the team knows that it is possible to send CAD designs over a network the resulting process design is unlikely to be very good. This is true but weakly put.

'Very well', the response may be, 'We will have people in our team who know that designs can be transmitted over a network and know other facts of that sort. That isn't asking very much. Hundreds of thousands of managers and consultants have a rough notion that such a thing is possible. However, let us be clear; we certainly don't need people with more detailed knowledge, about (say) the error correction procedures incorporated in IGES, one of the industry graphics standards.'

There is a confusion here. On the one hand, the detail of the IGES graphics standard is certainly not the kind of technology knowledge needed for process design work. But, on the other, merely knowing that CAD designs *can* be sent over a network is nowhere near enough. The truly valuable knowledge, making a difference between perceptive and naive process design, lies between those two levels. That is the starting-point for the thought-provoking part of the debate.

Continue the example of CAD for a product-development process. A database should probably keep track of multiple versions of parts and sub-assemblies that may need to fit together. You may need to go back to yesterday's version of sub-assembly A, and back to last Friday's version of sub-assembly B, while keeping the latest versions of all the other sub-assemblies. After seeing how these fit together, you may ask the system to substitute still other versions of the designs of certain parts, and so on.

Knowing the technology well enough to be aware that a multi-version facility *can* be required and *can* be met by available software technology is at least a start, but a small one. Many other requirements can be imagined for juggling several versions of engineering drawings. The valuable contribution, that will help establish the real design options, is knowing what kind of 'versioning' features are relatively straightforward (thus low-cost, low-risk) with current technology, and what are relatively advanced (thus high-cost, high-risk, but, if you are sure you really need them, perhaps high-gain).

This example suggests how delicate the problem is: you need some people with enough technology knowledge to operate confidently and reliably at the level just described, but with the right mentality to bring non-technology people into the debate. Or, put another way, you need to organise your approach to process design to promote the interactions between supply and demand factors as sketched in the diagram.

IT Approach Options

Carr et al are surely right that the degree of IT-stress in re-engineering design should depend on the circumstances, but what are the main generic options available?

These authors give a vivid brief example. A car rental company designed a new system to collect data about vehicles; this required a higher degree of automation; however, IT people were not involved in the main design work. After signing off the new design, one manager purred smugly 'This is almost as good as putting a computer in the parking lot.' An IT expert said 'We can do that, too, you know.' The revelation caused the company to

Demand for Business Features - and Technology Supply

Demand-driven, ie technology-independent, approach

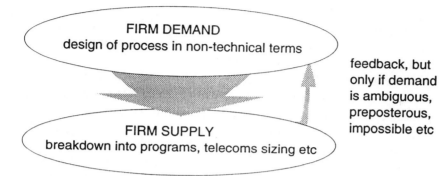

FIRM DEMAND
design of process in non-technical terms

FIRM SUPPLY
breakdown into programs, telecoms sizing etc

feedback, but
only if demand
is ambiguous,
preposterous,
impossible etc

Problem: this is choosing from a menu without any prices.
Therefore unlikely to produce the 'best buy' process design.

Supply-demand interaction approach

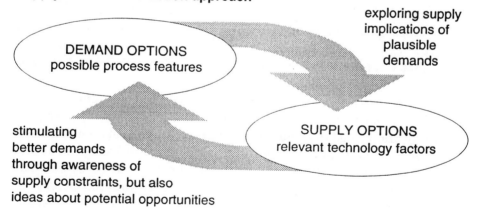

DEMAND OPTIONS
possible process features

exploring supply
implications of
plausible
demands

SUPPLY OPTIONS
relevant technology factors

stimulating
better demands
through awareness of
supply constraints, but also
ideas about potential opportunities

Aim: best-buy match of supply and demand.
Main problem: difficult to do!

scrap the new design and produce a better one, with hand-held data collection devices and wireless telecoms. Much of the earlier work was wasted, but at least it was only design work; had the chance comment not been overheard, the company might have invested in a greatly inferior system.

Reliance on this kind of arbitrary feedback must be unacceptable, but there is no one solution to the general problem. Here is a range of possible approach options:

● **Gradation 1.** The process design is made in non-technical terms, without input of technology knowledge. Once firmed up, it passes to the IT people who, in effect, translate it into a technology design — without coming back with any new ideas. After the anecdote above, this may sound foolish, but need not necessarily be so. If the process is (say) summarised management information, a person knowledgeable about technology might make a reliable judgement at the very start that the technology means were bound to be straightforward, and there was just no possibility of missed chances like the parking-lot computer.

● **Gradation 2.** Most of the design work is done in non-technical terms, but an outer ring of part-time team members with IT knowledge can be consulted from time to time. These people aren't expected to take a strong, active role, by suggesting (say) that each parking-lot attendant be given a hand-held computer. If the that kind of initiative is desirable one of the other gradations is more appropriate.

● **Gradation 3.** The design is developed in non-technical terms, but at least some of the people involved possess enough technology knowledge to avoid blunders and spot natural opportunities. Given this, together with the fact that the process doesn't raise any exacting technology issues, the design can be confidently firmed up and then passed on for translation into technology.

● **Gradation 4.** The design is made in non-technical terms, without the input of technology knowledge. Once it is complete and apparently ready, there is a separate brainstorming activity to find any missed technology-based chances, and if necessary, to rework the design, perhaps substantially. This brainstorming is a major activity, not just a check for obvious blunders: it might be budgeted at 10-20% of the effort of the whole design project. This is a kind of compromise approach for situations where IT seems more of a force than with the previous gradations.

● **Gradation 5.** IT issues constantly come up and influence the appraisal of design possibilities, and the team contains knowledgeable people. This approach is probably necessary for the CAD example discussed earlier, and perhaps for the parking-lot computer example.

● **Gradation 6.** As with the Davenport and Short proposal, one separate, early, relatively large chunk of the design project is explicitly devoted to using IT as a driver for innovative ideas. This could suit the parking-lot computer example, particularly if a variety of other technologies were also worth consideration.

This six-item menu should help in determining the right approach for any particular project. The corollary is that any standard methodology with just one approach for all situations, whether it be IT-free or IT-dominated, must be unsound.

▼ This argument against standard methodology applies on other issues in other briefings too — wherever the choice of best approach depends on the situation, ie almost everywhere. ▲

REPRESENTATIVE IDEAS

Hardly any of the authors in the bibliography take on the *significant technology* issue. Many mention particular technologies, but without demonstrating that they are specially influential in re-engineering.

Davenport's book is certainly an exception, but it is disappointingly literal. Dozens of uses of technology are illustrated as ways of supporting dozens of types of innovative redesign; but there are few powerful generalisations, such as 'technology X is found in example after example of re-engineering; technology Y is often vital for processes of a certain kind; and technology Z, though sometimes applicable, is nowhere near as influential as technology X.'

▼ Some statements of that sort might be dangerously crude, but more subtle, carefully qualified, generalisations could be made. The point is that generic *analysis* of how technologies and uses can be matched together has far more utility than mere processions of examples. The preference for literal, specific examples over careful, generic analysis is a shortcoming in much writing about management topics. ▲

DISCUSSION

One good tool for analytical insight here is the distinction between the four main kinds of processes that a re-engineering project may tackle.

Kinds of Data

To start with distinguish between several kinds of data:

● Traditionally, and still, most computer systems for the administration of an organisation deal with **mainstream** data: discrete, atomic data items, such as *customer-number* and *customer-name* and *quantity-ordered*, are structured into records (or tables or objects; it doesn't matter to this discussion) that are in turn arranged in a tightly structured database.

● **Non-mainstream** data, less widely used in administrative systems, covers everything else: texts, such as reports, laws, abstracts of articles, minutes of meetings etc; photographic images; spatial data, the data of technical drawings and maps; voice data.

Mainstream data is, by its nature, strongly structured. Non-mainstream data is more diverse: though spatial data (eg the technical drawing of an aeroplane) is strongly structured, a memo containing somebody's comments on a certain drawing is not. This gives the following possibilities:

● **Mainstream data**: of necessity, well structured;

● **Non-mainstream data, strongly structured**: notably spatial data, whose two main varieties are CAD/CAM (for technical drawings) and geographic (for drawing maps);

● **Non-mainstream data, weakly or subtly structured**: notably text; there are various ways of organising text, but given the inherent richness and imprecision of language, even the most rigorous won't be as strongly structured as the previous two categories;

● **Non-mainstream data, unstructured**: eg photographic images, images of documents, phone messages.

This doesn't mean that every application system uses one and only one kind of data. Hybrid databases occur: a database containing images of houses for sale may contain also mainstream, factual data and descriptive texts composed by the estate agent.

Kinds of Process, Kinds of Data

These categories of data map onto the kinds of process, though in quite a complex way. Here are some notes:

● Many **transaction-based processes**, whether before or after re-engineering, use nothing but mainstream data. The main exceptions are processes using document-images.

● The database for **matter-based processes** is more likely to be a moderate hybrid; eg mainstream and spatial data for the semi-standard products at AT&T; and, for an employee dismissal process, mainstream (personnel records) and text (legal depositions) data.

● Some **project-based processes** — notably those concerned with designing complex, technology-based products — use non-mainstream, but strongly structured data. Other project-based processes use non-mainstream data, that is weakly or subtly structured, together with other kinds of data too. For example, an advertising agency planning a marketing campaign may need mainstream data (eg demographic statistics), spatial (ie geographic) data, text data (eg consumers' and researchers' comments in a test market) and photos (for use in adverts).

● There are two long-established varieties of **facility-based process**: the decision-support system (DSS), that summarises large quantities of mainstream data, such as sales transactions; and the text database system, providing access to (say) an immense body of legal texts. Attractive ideas for a new or better facility-based process often bring in a hybrid database; eg storing texts and graphics along with mainstream data in a DSS.

What is the use of the account so far? First, it may help in suggesting design options. In a project-based process essentially concerned with spatial data, such as at Kodak or McDonnell Douglas, one idea might be to shift emphasis a little towards facilities for exchanging weakly structured or unstructured data; eg e-mail or voice mail or documents in a text database accessible through keywords. In a matter-based process, such as Hallmark — also product development, but without spatial data — a plausible idea might be to try and make data more structured; eg storing images of Christmas card designs, indexed by mainstream and text data, to be more accessible.

Second, the nature of the data can affect the way a new process is developed. If the process is based on highly structured data, then a large part of the effort may be to prove that everything does fit together perfectly without anomalies or inconsistencies. If most of the data is non-mainstream, weakly or subtly structured, then that consideration is less pertinent; here there is more need for an iterative approach to discover a design that is not too crude or feeble, but not too laborious or puzzling to use either.

Kinds of Process, Significant Technologies

Here is a summary breakdown identifying broad technology resources for achieving the goals of re-engineering in *transaction-based* processes:

● Take advantage of the maturity of large-scale **mainstream database** and **telecoms** technologies, as in the Ford example and many others that streamline well-structured administrative processes.

- Use other well-established technology appropriate for supporting workers in much broader roles. The low-cost, **networked PC** is the obvious example. **Expert system**, as at Mutual Benefit, is another influential technology here. **Document imaging**, is another, although its role is often misrepresented. In the office of the Ontario PM, the new process stores the image of a letter received from a member of the public, but the really crucial point is that members of staff can key in their own comments to go with the letter as an electronic file (as opposed to composing a memo, printing it out, making a copy, stapling it to a copy of the letter etc).

- Use the relatively new **workflow** software, that applies rules to control the flow of information between workers, and thus 'automate the procedure manual' — or rather a rewritten one. The challenge this technology tries to meet is to streamline and rationalise processes, while in some ways making them more subtle; eg coping with intricate exception conditions, and allowing complex feedback between people.

- Use any **other** form of technology not included in the above; eg barcoding or point-of-sale equipment, speech recognition, voice mail, neural network, etc.

It is harder to generalise about the technology for *matter-based* and *project-based* processes:

- **Telecoms** technology almost always comes in (though not invariably: Hallmark poets and artists were brought together in the same office). You may need more sophisticated telecoms than for a transaction-based process: transmitting colour images of cross-sections of an aircraft may call for more resources than sending the data of a thousand sales orders.

- The same point is true of database technology. **Non-mainstream database** technology is in general a more demanding branch than mainstream, at least in the following sense. With non-mainstream database there can be much greater variation between an ingenious and just a humdrum design, and there are far fewer people to be found with adequate skills and knowledge.

- Workflow software with its stress on predefined general rules to deal with masses of cases is less pertinent to matter-based and project-based processes, but the new **groupware** software is often applicable. Groupware helps link up technology elements and systems that are essentially separate things: word processing, spreadsheet, access to central or external databases, e-mail, scheduling meetings etc.

● **Other** technologies can be applicable; eg Hallmark's management process relies on data from point-of-sale devices.

REPRESENTATIVE IDEAS

Davenport's book scores again by being almost the only one to raise a topic affecting many re-engineering projects: the possible *constraints of IT* on re-engineering.

IT Constraints

Davenport explains lucidly, honestly and incontrovertibly that IT can bring constraints as well as opportunities. It may be just silly for a re-engineering team to assume that it has a free hand to design a completely new process, without reference to existing IT systems. For example:

● The new design may entail large investments in new IT systems to replace the existing ones. If these were acquired recently at large cost at great cost, managers may be unwilling to countenance such a drastic step.

● The new design may integrate the activities of separate departments more tightly than before, but many of them may have separate, incompatible computer systems. If so, the cost of replacement or modification may be unacceptable.

● The new design may alter the interactions with customers and suppliers. but it may depend on them altering their IT-based systems. Experience and common sense suggest that, whether for good or bad reasons, they may not be prepared to do that. The Baxter firm has had six generations of order-management system, each innovative and bringing steadily greater integration, and yet some customers still prefer to stick with the second- or third-generation systems.

As Davenport points out, it is surely better rationally to recognise such constraints, their implications, and the possible tradeoffs, than to ignore reality and produce designs that will never be successfully implemented.

DISCUSSION

As on several other issues Davenport is a lone voice in pointing out something of great importance, and yet he doesn't drive home its full implications. For example, he doesn't go as far as explaining the unavoidable conflict between IT infrastructure — planned at organisation-level for cost-effective shared use — and radical re-engineering, focused on one part of the organisation, and producing large-scale, unexpected innovations.

Example of the Infrastructure Problem

Since the subject of IT infrastructure is rarely associated with re-engineering, the first step is to see the problem. The following diagram gives a general impression, and the text of this section reasons things through.

Suppose, as in Hammer and Champy's example, that Imperial Insurance is re-engineering the claims process. The existing process is a computer application, and probably makes use of a database, that, in one coherent structure, contains data about many parts of the company, applicable to many processes. Moreover, the whole configuration of PCs, terminals, computers, telecoms lines and associated technology is planned to support the whole workload of the company's application systems, not just claims processing.

There would be no difficulty, if it were possible neatly to remove all pieces of this infrastructure that concerned claims, and then replace them with new pieces corresponding to the

Infrastructure: the General Problem

Concept
Design everything as a general structure, organisation-wide;
so that, later, detailed development work can be fitted in conveniently.

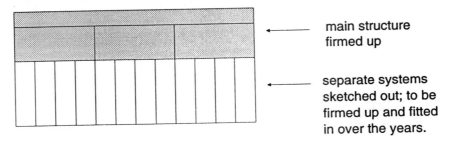

←——— main structure
firmed up

←——— separate systems
sketched out; to be
firmed up and fitted
in over the years.

Problem
Suppose your redesigned processes turn out not to fit that structure.

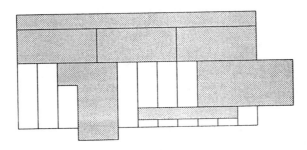

redesigned process. But it is probably not feasible (or if feasible, disadvantageous) to do this. Here are some example problems:

● **Logical repercussions.** The database (for the sake of example) is structured on the basis that for each insurance-holder there may be several policies, and for each policy several claims, and for each claim several payments. Certain codes are held at the level of insurance-holder (eg a profession-code, since accountants are thought to be safer drivers than popstars), while other codes are at the level of policy (some types of car being inherently more risky than others).

Suppose one innovative feature of the new claims process is an intricate piece of logic to decide which claimants are likely to be trustworthy and which fraudulent, and, in order to work, this feature requires new coding systems at the insurance-holder level. But these codes are used by other applications besides claims. It may not be possible to make such changes to the database design without ramifications for the non-claims applications too.

● **Physical repercussions.** People in the claims department also have general-purpose office-automation needs, eg electronic mail and word processing for letters and reports. At the moment (for the sake of example), the same hardware infrastructure is used for this as for the claims process proper. Presumably that should continue after re-engineering. But the redesigned claims process (very likely) has a different pattern of processing and data transmission from the old: maybe a lower quantity of longer messages, with a different pattern of peaks and troughs. Will this hardware infrastructure (or some simple modification of it) handle the new claims process efficiently?

To decide that, an expert needs to study the requirements of both the new claims process (reasonably enough), but also office automation (which is meant to be outside the scope of the re-engineering project). Worse, the change in workload from the claims department could also affect the data centre, where the central database of the whole company is held. Perhaps the new processing and transmission requirements cross some threshold, so that massive upgrades of mainframe computer and telecoms equipment are triggered.

Logical and Physical Infrastructure Ripples

As these examples suggest, there are two main complications: ripples of *logical* implications, usually affecting database structure and content; and ripples of *physical* implications, usually affecting the capacity and power of hardware.

You can scarcely ignore such implications, but following the ripples may necessitate so much work outside the defined scope of the re-engineering project that things get out of control, or even never-ending. If study of claims process requirements leads to study of office automation, then it makes sense to allow for likely developments there too — more advanced desktop publishing or perhaps groupware. But if innovations in office automation are being studied in the claims department, shouldn't they be studied in other departments too?

It may be objected that a truly methodical and logical person would first set up an organisation-wide infrastructure, intended to be good in its general outlines for the next five or, better, seven years. In this environment individual projects for individual processes could be born and develop, benefiting from the infrastructure, but not changing it.

During the eighties this was a very popular but, all in all, pretty unsuccessful concept. It rests on the fallacy of perfect knowledge. If, at the time you design the infrastructure, you can know everything relevant to sound long-term design decisions, then of course the approach is desirable. In practice, in most situations, there are too many uncertainties for detailed, top-down planning of logical and physical infrastructure to have much lasting value. You can't know now the characteristics of the innovative application systems that you will authorise for development in three years time (if you could, they would not be innovative). Therefore you can't know what kind of infrastructure you should have at that time. Re-engineering, with its stress on bold, convention-breaking, radical redesign of individual processes, delivers the *coup de grace* to detailed top-down planning of IT infrastructure.

Tackling the Infrastructure Problem

This is a problem with no entirely satisfactory solution. Two desiderata conflict:

● In making changes to logical or physical infrastructure necessitated by a new process design, it is desirable to leave no major loose ends, make no great over-simplified or arbitrary assumptions, and leave out no pertinent factors.

● On the other hand, it is also desirable to carry out a re-engineering project with a well defined scope, without excursions beyond its boundaries into the enterprise-wide infrastructure.

These two being often impossible to reconcile completely, the most basic piece of advice is at least to recognise that there is a tradeoff challenge here. Although this infrastructure tradeoff arises time after time to some degree, the degree does vary greatly. Some conscious assessment is desirable of how formidable the difficulty is in the particular instance, and whether the awkward implications are logical or physical or both:

● The example above of implications for logical infrastructure is nowhere near the most messy conceivable. Suppose, instead, that the design team proposes that the assessment of any claim should be influenced by information in the database about the policies and claims of the claimant's spouse and relatives (even with different surnames), and of any businesses the claimant owns or works for. Overhauling the database structure to build in the necessary links for this purpose would not be trivial or limited.

● If, in the claims department, the ratio of claims-handling work to general office-automation work is (say) 30:1, and if, at the data centre, claims processing is only 5% of the load, then the chances of messy implications for physical infrastructure are low. But if the numbers are very different, then the chances are much higher.

With this background, it becomes clear that there are three possible approaches, albeit with many situation-specific variants:

● Agree small, piecemeal changes to the infrastructure with whoever is responsible for it (eg the data administration department), explicitly documenting assumptions and implications.

● Recognise that there are substantial implications for infrastructure, but that they are controllable: they need not spread like wildfire to raise issues and exact decisions, that in turn raise new issues in other parts of the business. Set up a formal structure for interaction between the re-engineering team and the guardians of the logical and physical infrastructure.

● Recognise that the process to be re-engineered is so mixed up, in such complex ways, with infrastructure issues, and thus with other processes, that the definition of scope is not a tenable one for a coherent design. Go back and redefine the scope.

CONNECTIONS

The importance given to IT is one variable in differentiating people's attitudes towards re-engineering. Some but not all see it as a central, defining feature. Were that view to be general, it would have implications for the very definition of re-engineering, discussed in Briefing 2.

The subject of infrastructure has obvious associations with what Briefing 3 calls the exhaustive strategy: attempting to re-engineer everything. In both cases the essential difficulty is to set up a co-ordinating master blueprint that is organisation-wide, and to maintain it in the face of innovative developments on individual projects.

9. Process Design Decisions

This briefing discusses some topics that may at first seem disparate, but are in fact closely connected.

Design Principles and Megatrend Theories

Many books put forward some design principles for re-engineered processes. 'Eliminate review tasks', 'push work up, not down', and 'organise work groups around extended processes, not tasks'; these are typical tenets from the book by Petrozzo and Stepper.

The trouble is that such advice, when examined carefully, often turns out to be imprecise, or, if precise, implausible or so heavily qualified as to be of little interest. For example, 'eliminate review tasks' surely isn't acceptable if taken literally as 'eliminate all tasks whatsoever that review whether work has been done correctly, irrespective of any fatalities, nuclear fallout or other disastrous consequences that may ensue.' But if, on the other hand, it is to stand for 'eliminate *unnecessary* review tasks', then it is empty as advice because nobody would disagree with it.

The principle might be interpreted as something more subtle, such as 'Always design a process so that all errors and imperfections, however minor, are caught as early as possible — thus eliminating the need for large tasks near the end whose sole purpose is to carry out checks.' But then why should that be universally valid advice? It means that you should never design a process with a control procedure at month-end to spot any discrepancies that have arisen among thousands of cases; you should always ensure that, however much extra work is involved in handling each case, any discrepancies will be spotted immediately they arise. Why be so dogmatic? Surely there can be situations where the former choice is the better buy.

Though some general design principles seem to have no validity, however interpreted, it still seems cavalier to sweep away all such material as worthless. How then can the fruitful advice be sifted from the rest, and applied to design better processes?

A similar question arises with megatrend theories. There are some rather abstract theories about re-engineering expressing a new way (paradigm, even) of organising work: it exemplifies a new form of industrial engineering, is associated with the knowledge-based economy, represents a rejection of the principles of the scientific school of management, and so forth. But does this kind of thing have any value? What reason could any manager in an organisation have for caring about it? Perhaps a megatrend theory can help in deciding better how to design or implement a re-engineered process. But is that actually true?

The most natural answer to all the questions so far is that certain design principles and general theories, if they offer more than platitude and excessive abstraction, may help in exposing the *design options* open for any new process. Conversely, material that fails to further decision-making in this way probably isn't worth knowing about.

The Importance of Options

Re-engineering authorities say amazingly little about options, and the standard methods of many consultants are little better. What is so vital about design options?

First, to decide anything is to choose one option rather than some others. If you decide to do X, and there was no other possible option, then you have scarcely taken a decision at all. Or if there were many plausible non-X possibilities that you never even considered, then it is quite possible that your choice of X is a bad one.

It is rather unsatisfactory to present a design for a re-engineered process with the message: 'Here is the product of our work; either accept it, or reject it and stay with the old way of doing things.' Any shrewd participant in decision-making is entitled to respond: 'If you show me no other options, how do I know that that this design is on balance the best buy? Perhaps an alternative design, more limited, but at half the price and more quickly implemented, is a better buy; or perhaps some more expensive choice using IT in a more adventurous way has much to be said for it. If you have done the work to evaluate other options, please describe them briefly with their pros and cons; then if I do accept your recommendation, I will know what I am rejecting. If you haven't evaluated any other options, how can any of us be confident that the design you propose is in fact the best?'

How, in general, should you generate and make sense of significantly different options affecting the main shape of the process? One key is to recognise the notion of *tradeoffs*.

Suppose everyone concerned with a certain process agrees that factor A (cycle-time, for example) is so weighty that nothing else matters. Then the goal to attain is the design that optimises A. Working this out may not need much debate about starkly contrasted options. But such situations are exceptional. Often A, B and C are all material factors; one possible design is very strong on A, moderate on B, and very poor on C; another option is moderate on all three; another has a different balance of merits; and so on . . . Here searching out design options that provide different balances of tradeoff factors, and working to refine the most pertinent of them are fruitful activities.

REPRESENTATIVE IDEAS

Quite a lot has been published about the design principles that characterise successful re-engineered processes. This section is largely concerned with imposing some order on that material.

The anti-Taylorism Argument

Beneath all the hype the early part of Hammer and Champy's book does contain a respectable piece of argumentation, which (in entirely different words) can be put like this:

● Any organisation has to break its operations down into discrete but related pieces done by different people; a motor manufacturer where each one of 10,000 mechanics also made sales to dealers and negotiated contracts with suppliers would be chaotic. Some degree of specialisation of role is unavoidable in any substantial organisation at all. A factory's production line with each worker endlessly repeating one small task is an extreme instance of this principle.

● Traditionally, the *administration* of business has been organised analogously to production in a factory. Hundreds of clerks each perform specialised, limited tasks, that in combination make up one intricate system.

● The best possible way of organising a factory, office or anything else is not necessarily the one that fragments the work and specialises the roles to the maximum. The more fragmented the work is made, the greater the costs of co-ordinating it all. These costs include the employment of layers of supervisors and managers, extra work done merely to plan and to check, time wasted in queues between fragmented operations, and a disproportionate effort to sort out a minority of cases with non-standard features or hitches between operations. Thus the real challenge is to find the right balance, *trading off* the efficiency and discipline of fragmentation of work against the costs of co-ordination in a highly fragmented system.

● This tradeoff challenge has held for many decades, varying from firm to firm, from time to time. But now, in the nineties, the optimum balance for many organisations (in office work, not necessarily in factories) has become less fragmented, and gives people wider roles, and reduces the bureaucracy and paperwork of co-ordination. Why? Because now there is no need to fill in five forms and send them to five people, each with a specialised task; supported by information technology, one person can do the work the five used to do. IT reduces the costs of co-ordination. In short, 'When a company reengineers, once complex processes become simpler while once simple jobs grow complex.'

● And there may be softer benefits too: people playing wider, more stimulating roles may be better motivated, may contribute to continuous improvements, and may perform better in unpredictable circumstances.

Hammer and Champy don't use the term Taylorism but it often crops up with other authors who make much the same argument. Frederick Taylor was a management guru at the beginning of the century who spread the principles of extreme task decomposition and specialisation of role, particularly in factories. Today nobody has a good word for Taylorism; the term has come to stand for an authoritarian, inhuman style of management.

▼ The stress on tradeoffs in the argumentation above is not present in Hammer and Champy's text. That claims boldly to be reversing the industrial revolution, rejecting division of labour and economies of scale, etc. Read quickly, it might even be taken to suggest that the less fragmented any system the better, and the wider the roles the better — without any qualification at all. This, of course, would be nonsensical.

A handy tool of critical thinking is the (condescendingly named) *principle of charity*: before getting into somebody else's argument, express it as cogently as you reasonably can; then you can consider what is most striking about it, without wasting time on tedious imperfections. Hammer and Champy's argumentation is far more plausible and worth thinking about when the bombast is hacked away and the logic is expressed in terms of tradeoffs. ▲

Generic Design Concepts and Examples

An article by Hall et al features the example of BAI, an Italian bank. A customer's cheque-deposit transaction used to involve 64 activities, 9 forms and 14 accounts; after re-engineering, only 25 activities, 2 forms and 2 accounts were needed. The personnel in each branch has been halved, and 50 new branches have been opened with no increase in staff overall. This is an example of an empty example: with such elephantine procedures to start with, it may be safely assumed, anybody selected at random from the street could have come in and suggested a better process design by sheer common sense. The account of BAI's success doesn't tell you anything of general force that will help you design your own process better. Other case-studies can be criticised for the same reason, but some do provide genuine insights into the principles of re-engineering.

Hammer and Champy's IBM Credit had a deeply Taylorist, multi-step system for the processing of new requests for credit to buy computers. The average cycle-time was seven days, while the actual work done on the case, after all the fragments were added together, amounted to only 90 minutes. Another shortcoming was that no one person knew the status of the request, ie its current position in the multi-step process. With the new re-engineered process there are fewer operations, and one person, a deal structurer, is responsible for virtually all of them. Now average cycle-time is only four hours, with no increase in staff numbers, even though the number of requests handled has increased by a factor of 100 (yes, 10,000%; however, there is no claim that the dollar value of the deals has increased by this factor).

Again, it might be argued, if the old process was that poor, even the dullest trainee analyst could have devised a better one. True, but the interest in this example goes deeper than the figures. Two specific design concepts are illustrated:

● If the jobs of people are widened far enough, the concept of the **case worker** (here called a deal structurer) opens up: someone responsible for all the operations in one case. This has obvious benefits in both customer service and staff motivation.

● The old process was slow because it was complicated, and it was complicated because for certain tricky deals complicated procedures are unavoidable. But the tricky ones are a minority. If the tricky are separated from the rest, the majority can follow a streamlined path — and average cycle-time will be reduced. This concept is called **triage**, after the French practice of sorting (*trier*, to sort) casualties in overworked battlefield clearing stations into three categories: will die anyway, spend no time on; should survive, no need for immediate attention; and immediate attention will make all the difference.

These are two tangible, general design concepts that can help in making sense of case-studies in journals, and in getting a grip on real-life situations. Among other well-known examples of re-engineering, some show the case-worker concept, some triage, some both, some neither. One of the most notable features of the famous procurement process at Ford is that, as far as the published accounts go, neither the case-worker nor the triage concept is at all prominent.

▼ This is not to criticise, merely to point out that different themes appear in different examples to different degrees. Therefore, it seems plausible, a good way of finding the best process design for a certain process, may be to think out possible design options that also express these themes to different degrees.

The general technique of recognising characteristic but not compulsory themes can help in all kinds of design work: designing an electronic data interchange system or quality assurance procedures or a strategic-alliance agreement, for example. ▲

More Generic Design Concepts

Triage and the case worker are only two of the themes to be extracted from examples. Some authorities present long lists, but usually with many points saying much the same in different words. Here are six promising general principles of design, selected out of all those mentioned by Hammer and Champy:

● Combine what were previously several jobs in one person; thus leading towards, but not necessarily going as far as, the concept of the **case worker**.

● Use **triage** to have in effect several sub-versions of a process.

● Get several operations on a case done in **parallel**, rather than strict logical sequence. Bell Atlantic used to collect all data conceivably relevant to billing a customer, and only after that start planning the installation of a telephone line; now these things happen in parallel.

● Allow and encourage **decision-making by workers** rather than managers. This will tend to eliminate some managers' jobs and remove some levels of hierarchy.

● Shift the **boundary of responsibilities** between business partners. The supplier of Pampers disposable nappies is allowed to make decisions on its own initiative about what to deliver to Wal-Mart and when.

● Have fewer, more carefully chosen, **checks and controls**. It isn't necessarily cost-effective to track down tiny discrepancies or instances of dishonesty immediately they occur, or even at all. A better buy may be a simpler process to check for such things at month-end. It may not be necessary to check thoroughly everything a business partner does, eg every garage invoice submitted for an insurance claim. A neater policy may be to develop long-term relationships with trusted partners, and let it be known that any dishonesty discovered will dissolve the relationship.

One popular society-wide megatrend is, though present, somewhat understated in the literature of re-engineering. Most people probably feel that giving individuals broader, richer and more interesting jobs is a good thing in itself — and therefore worth doing even if all cost and other factors balance out completely. But should a process design option that is strong in job enrichment ever be preferred over another option weaker in this factor, but stronger in (say) cost-effectiveness? It might be expected that a minority of idealistic management authorities would go this far. However, hardly anybody does.

DISCUSSION

Can the material about design principles and megatrend theories be used to generate options and thus informed design choices? This section tackles that question.

Distinguishing between the
Taylorist Weaknesses

The tradeoffs between the pros and cons inherent in the anti-Taylorism argument deserve more careful study than Hammer and Champy provide. Their many bad examples of excessive fragmentation contain a great variety of possible disadvantages. The following table lists the main ones. Contrary to the impression given by some polemics, it isn't the case that all systems from the era before re-engineering suffer from *all* these shortcomings. This list of typical weaknesses can be used as an intellectual tool to help focus on those options that matter in any particular instance. For example:

● 'Design option A for the new process focuses primarily on solving the queueing problem, which, we judge, is much the main challenge in this process. By concentrating on that one thing it brings large benefits at low cost.'

● 'Though queueing is a big factor, so are flexibility and initiative in special situations. But the design option that has the shortest queues and thus fastest throughput *on average* turns out to perform rather unsatisfactorily at peak periods such as Christmas or, worse still, emergencies such as a postal strike. Design option B is more complicated than A, and contains somewhat more queueing, but provides more opportunity to control peaks and emergencies.'

● 'Not all the eleven weaknesses listed apply to this process, but many do, albeit to varying degrees. Design option C, the most ambitious, tackles the main ones as forcefully as possible.'

▼ Most organisations should not care very much whether they participate in the anti-Taylorist, or any other, megatrend. What counts is taking the best decisions for the particular organisation in the circumstances. Starting out from general statements about the state of the world and ending up, through a sturdy chain of reasoning, with 'therefore our new process should be like this . . .' is rarely a sensible aspiration. This principle holds in management affairs far beyond re-engineering. ▲

Highly Fragmented Systems: Typical Weaknesses

The costs of managers and supervisors who merely control, rather than do anything useful, are unproductive overheads.

A tremendous and disproportionate effort may be needed to put right relatively small problems and inconsistencies.

Incompatible department objectives may lead to suboptimal results for the whole organisation; eg the performance of department A is assessed on its speed of meeting customer orders, but it is handicapped by dependence on department B, which is assessed on keeping down inventory costs.

Elapsed time may be wasted in queueing; if there are ten steps in a process, and each is handled by a different person, then there are also ten queues a case must wait in.

No one person may know or care about the status of (say) a customer order.

Staff may be demotivated by repetitive performance of limited tasks — a waste of human resource.

Cross-boundary inefficiency may arise, eg one manager won't allow a technician to go immediately to another location to repair a machine, because of the cost of an over-night hotel; so the machine, which is owned by a different manager, is repaired a day later, at a cost to the whole organisation far greater than the hotel bill.

The work in even a small operation may be defined in tremendous detail, and thus become complex, since it is meant to enable a person of little initiative or knowledge to handle practically all possible cases, including very tricky ones.

Controls to catch errors or abuses may be so elaborate that they cost more than the minor errors they catch.

The limits of system complexity that is practically controllable may be reached even before certain desirable extra features can be built in.

Flexibility and initiative in special situations may be very limited, because of the system's effect on workers' psychology, and because of each person's limited knowledge of the whole.

Generic Design Concepts as Source of Options

Generic design principles may be fascinating to read about, but how can they help anyone deal with a specific process better? In particular, can they be a shortcut to spotting design options?

It is tempting to conclude that all the design principles are nothing but variations on the principle of Occam's Razor: choose a neat, simple way over a complex one — unless there be some good reason for choosing the complex one. Nevertheless, a carefully chosen armoury of design principles can be valuable. For example, the principle of *triage* can easily stimulate option-generating trains of thought:

• 'Maybe for the order fulfilment process in our wholesale business the design should include a simple two-way triage near the start: simple orders (for a main-line product) and all other orders. That common-sense measure allows a superior new process to be set up, without undertaking anything more ambitious, such as a special investment in technology.'

• 'But another design option is to have a more complex process with two-, three- or four-way triage occurring at several branching points. Orders for main-line products would be separated from orders for special or non-standard products; but, for example, orders for exceptionally large quantities would take a different path; as would orders that need not be delivered urgently and thus enjoy an extra discount. That would be more efficient than the simple two-stream option, at least in theory. But perhaps it would be over-complicated.'

• 'And another design option would avoid any strong triage at all. All orders would pass along the same stream. The characteristics that differentiate each order (type of product, quantity, urgency etc) would be allowed for by specially-developed expert system software. That would be the key to streamlining the process.'

More Examples of Option-forming

The examples so far apply most obviously to the transaction-based process, but in a matter-based process, with its relatively substantial case, the triage opportunities may be richer still.

For example, at a telecoms utility, the existing process for handling compensation claims and litigation with customers may include some crude triage already, into two or three categories, based on intuitive assessment of a case's complexity. One design option might be to increase the number of triage streams, but other options might arise from more subtle application of the triage principle. Perhaps four or five complexity factors (eg worst-case cost, possible PR-impact, one-off or endemic defect, etc), could be part of an algorithm to decide which triage stream any case should follow. The formula might even take account of how many other matters were already going through each stream, in order to achieve some balancing of load. Plainly, a number of design options of differing sophistication are possible.

Again, with the case-worker principle, it is wrong to reduce the choice to applying or not applying the concept; there may be several ways or using it that lead to plausible design options. For one thing, there is a generic choice between the concepts of the *case worker* (who actually does most of the work on a case) and the *case manager* (responsible for ensuring that the work is done, though not for actually doing it all, and constantly in touch with the status of the case). One design option might implement the pure case-worker concept; an alternative the case-manager. The same mode of thought can be applied to the other design concepts listed above.

It does seem worthwhile to foster the habit of looking out for opportunities to apply variations on these generic design concepts. They can become an almost instinctive tool for finding the options available in any situation. It may seem tempting to add in a few more design concepts too, but eight or twelve or fifteen are by no means better than six. The larger the set, the more likely it is to contain overlaps, and thus hamper incisive thought.

REPRESENTATIVE IDEAS

Many authorities admit that there is not much to say about how to find an innovative design for a process. But this concession is too great; the topic of finding and contrasting design options cries out for further examination.

Rare Example of Design Options

Michael W Dale in the book edited by Spurr et al gives a rare example of a choice between two design options. A manufacturer of electronic sub-assemblies is re-engineering the process of turning enquiries from customers into firm tenders based on costings.

Dale describes one straightforward strategy (option A in the table headed *Four Options*), and one other, more glamorous, alternative (option C, or close to it). He makes the valid point that substantial benefits can often come without much new use of IT, and the fanciest process imaginable is not necessarily the best buy. But this example will surely stimulate an alert reader to see more than the two options mentioned. The table suggests four, and it still leaves some other areas for choice untouched. Here are two:

● If the product is one with numerous customisable choices, then one possible refinement is to have an expert system to guide customer, sales rep or office staff to find the most sensible configuration for the customer requirements. Thus there are two versions of option A (with or without this feature) and the same applies to options C and D.

● As well as just working out the details (price, delivery date etc) of the tender to the customer, a process might also generate a proposal *document*, choosing between a variety of standard texts (eg descriptions of product features, contractual conditions etc) in an intelligent way. This could be relatively straightforward, and therefore worth doing as part of any of the other options, or perhaps immensely complicated and part of some higher option.

Four Options
Re-engineering the Process of Tendering for Customer Orders

Option A   Use common sense to streamline the present cumbersome office procedures: triage for different types of tender; people from different departments brought together in same office; etc.
No special use of IT.

Option B   Give portable PCs (with necessary software and data) to sales reps for immediate tendering out in the field — but only for the simplest cases.
For other cases the portable PCs send data about customer requirements to HQ, where the office work is done in the style of option A.

Option C   As with B, field sales reps give tenders almost immediately — but for most cases, ie all but very complex or large ones.
The portable PCs use more elaborate software than with B, and are loaded with more data — including, once a week, the factory's production schedule.

Option D   As C, but the portable PCs interact directly with production planning systems back at HQ (instead of being updated with a schedule once a week).
Thus cost and time estimates are based on more accurate information. Production capacity can be reserved even while the quote is being considered by the customer.

The design options suggested are far more than just differences of detail. They represent strategic choices calling for decision. It would be scandalous for an organisation to choose one option without giving some consideration to the others. And yet hardly any writer on re-engineering seems interested in identifying options like these.

▼ Keenan's article quotes Hammer's views on this same business process: preparing tenders for customers. There is the bad-old way with forms and two-week delays, and the good-new way, with portable PCs for sales reps and instant quotes; the bright manager will take the good-new way. This is an example of what might be called the *black-and-white fallacy*. In real life, there is hardly ever a clear choice between two main design options, one good and one bad. There is almost always a variety of credible options, each with a different balance of good and bad points. Ignore that, and decision-making becomes unacceptably coarse. ▲

Scraps of Ideas about Design Options

Morris and Brandon offer a rare standard methodology that explicitly recognises the notion of comparing options. But since they give no illustrations of the kind of options they have in mind, this major point goes for nothing.

Cross et al illustrate standard documents for defining options, but only blank forms, without filled-in examples. One summary sheet contains eight items, such as characteristics of the option or idea, potential obstacles to piloting the option or idea etc. An *options integration matrix* is a chart showing relations between general objectives and specific options considered. Thus (to invent an example) it would show that for a wholesaler, the design option 'deliver goods before checking credit-worthiness' is strongly related to the objective: 'improve customer satisfaction'; neutral to the objective 'enrich employees' jobs'; and contrary to the objective 'reduce bad debts'.

Davenport's book, unlike most others, strives to provide analyses and examples that will stimulate the reader to fruitful ideas. For example, there are nine generic categories of opportunity for supporting re-engineering with IT: automational, informational, sequential, tracking, analytical, geographical, integrative, intellectual, disintermediating. Examples are given of each. This is followed by a much more literal survey of 16 innovative applications: automated design systems, electronic markets systems etc. The last of the book's three parts consists almost entirely of specific examples of applications and technologies. This material seems intended as a prompt or checklist to generate ideas in any particular situation.

One drawback is that the categories tend to overlap. It may be an arbitrary choice whether to call a certain use of IT tracking (ie closely monitoring process status and objects) or analytical (ie improving analysis of information and decision making). One application may well have all the following traits mixed together: geographical (ie co-ordinating processes across distances); integrative (ie co-ordination between tasks and processes); and disintermediating (ie eliminating intermediaries from a process).

A graver shortcoming is that the material scarcely helps with the definition of rival design options. For example, Davenport gives textual composition as an application, and tells how a system can enable a sales representative to generate the text of a sales proposal semi-automatically. If this idea has never before occurred to you, the hint may be salutary, but the astute decision-maker needs to grasp how a great many *variations* of the idea can arise for consideration:

● One design option may pull together in one document standard texts corresponding to the type of customer and model of (say) combine-harvester ordered, leaving gaps and prompting for other more specific data; that is relatively simple.

● Another option that asks questions about the crops to be harvested and the type of soil, works out the appropriate selection of customisable options, and produces a complete draft proposal and contract is much more ambitious.

Many possibilities lie between these extremes. The trick is to set out the options and choose the most suitable one for the situation. Admittedly, to map out the main options within all the applications Davenport describes might take a whole book in itself. But to mention many innovative applications briefly is to promote the black-and-white fallacy: either adopt the innovative way described or stay with the existing methods.

DISCUSSION

Unless the example of the customer tender process is very unusual, there must be widely different options to be found in many other processes too. How should you go about finding them? Are there any general techniques?

Options Generated by Hard and Soft Factors

The crucial design options are those offering different balances of tradeoffs. Therefore a powerful way of generating options is to consider opposing factors that may be traded off.

Some tradeoff factors are *harder* than others: financial investment costs and savings in running costs are are among the hardest; but quality of customer service is a relatively *soft* factor. There is often a conflict between hard and soft factors. Though the panegyrist of innovation may not point this out, the design option that adds features to improve customer service (soft factor) may be less streamlined and more expensive (hard factor) than other options — particularly if the present level of service is recognised as too frugal. For example:

● Hammer and Champy describe how the **Whirlpool** company ensures that a customer's service calls are routed to the same representative each time. This must be a more expensive design choice than the option of a service-calls process without that feature.

● One article relates approvingly how the newly re-engineered process of one **insurance company** arranges for a representative to be informed of a household burglary, and to go out almost immediately — even in the middle of the night. At night, of course, the representative can offer no more than sympathy and the reassurance that an inventory of what has been stolen will be made as soon as daylight permits. Unless the company employs only saints there must be a cost to this facility. The redesigned process may well be the best buy on balance, but unless some less humane but cheaper option was also considered, nobody can be sure.

● It might seem from Hammer's influential article that the re-engineering at **Mutual Benefit Life** was immune from tradeoffs between hard and soft factors: hard cuts in cycle-time were achieved, and also soft job-enrichment for the new case workers with wider responsibility. But, on the account of this case-study in Davenport's book, things were not that simple at all. Many of the existing workers were judged not suitable, even with extensive training, as case workers. And this was in a firm with a tradition of hiring and training less-educated workers from the local community. The new process was set up at considerable cost to the company's paternalistic culture and to employee morale. In other words, a very awkward tradeoff choice was involved — or at least, should have been. There must surely have been several less radical options that offered different balances of improvement and upheaval.

Service Quality and Kinds of Process

As the above examples suggest, quality of service is one of the main generic soft factors. But there are gradations: reducing turnround on customer calls for service from an average of three days to three hours is a harder factor than ensuring that all service calls from a given customer are always handled by the same customer representative.

There is another angle: some fairly soft improvements in service quality are *extensions* to the basic service provided: eg Hammer and Champy's Imperial Insurance redesigns claims processing, and adds in new service features, such as visiting accident victims in hospital. This suggests some generic options:

● improve the service in **hard** terms (making it swifter, cheaper etc) without otherwise altering it; eg process an insurance claim quicker;

● provide much the same service with **soft** improvements; eg process an insurance claim in a more friendly, more informative way;

- provide an **extended** service, with new second- or third-order features; eg send representatives out in the middle of the night or to a hospital as part of the service.

This is not a simple three-way choice, but rather an account of three *aspects* of service quality that can stimulate contending design options. Thus, option A might aim for mainly hard, quantitative improvements; option D for an extended service without cost-savings; and options B and C for other balances of the various types of improvement.

Not all these generic options will apply in every situation. The *kind of process* can be pertinent in assessing which options really deserve consideration:

- With an extreme **transaction-based process**, such as insurance claims, it is relatively easy to separate out the hard from the soft factors, and thus sketch out options on the lines suggested.

- With extreme **project-based process**, such as Kodak's new product development, the distinction tends to break down. Plausible improvements to the process make it easy, instead of laborious, to check that two sub-assemblies which should fit together actually do; or to scrap one version of the design for a certain sub-assembly and return to a previous one; or to avoid mistakes in recognising and working through the wider implications of some change. Such things bring hard, quantitative improvement in that certain tasks within the design process will be done quicker. But they can equally be called soft improvements — making the designers' work easier, encouraging them to try out more variant possibilities, and therefore leading to a better-quality final design of the product. Thus for project-based processes, it is much less natural to use the different aspects of service quality to generate options.

- With a **matter-based process** (eg the building-permits at Rheden), recognising the three aspects of service quality may be quite a valuable stimulus. As well as the four typical options already sketched above, there is the more extreme possibility of *reducing* the soft quality aspects at the expense of the hard. The Rheden people might spend less time helping applicants to improve their permit applications, and use the resources thus freed to achieve improved cycle-times for defect-free applications. Or, again, they might go the opposite way and make their process far more akin to an iterative, creative, project-based one, where the distinction between the hard and the soft is blurred.

Kinds of Process, Streamlining and Enabling

The section above is mainly concerned with the effect of the process on the customer. Useful insights also come from looking more directly at what goes on inside the office. In a re-engineering project there are often two different, at times contradictory, goals (the diagram shows the distinction impressionistically):

● **Streamlining**: making the process simpler, neater etc;

● **Enabling**: making the process richer, ie enabling people to collaborate better, and have more and richer interactions; also enabling individuals to access more information, or the same information more easily.

Again, it is worthwhile to consider the kind of process:

● A **transaction-based process** has a definable sequence of steps, where the main variations may be complex, but are predictable. Here the natural (though not necessarily the only) goal of re-engineering is a slicker, crisper, more *streamlined* way of administration.

● In a **project-based process** the work isn't really administration, and it has no normal sequence of steps, and variations can't reasonably be defined very closely. Often too, the work is split between people working in parallel, interacting in intricate, unpredictable patterns. Here the natural goal is to *enable* people to carry out their work much more efficiently.

● In many **matter-based processes** *both goals* are attractive. This makes the distinction a fertile source of design options. You might consider a stark choice between concentrating on one goal or the other, and thus changing the character of the process; but other worthwhile options might mix both types of advantage.

● The **facility-based process** is essentially concerned with *enabling* the sharing of information in rich, unpredictable ways.

Streamlining and Enabling

Streamlining

Make a complex but definable administrative process simpler.

Replace this   By this

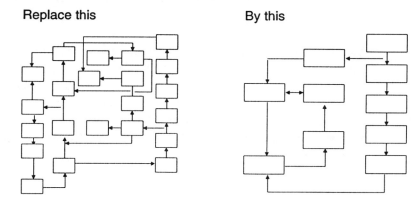

Enabling

Take a process where parties interact with each other repeatedly in sporadic, unpredictable ways, and make their interactions richer, faster, more effective.

Replace this   By this

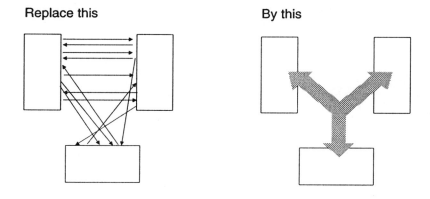

CONNECTIONS

This briefing is about the products delivered by the design activities: definitions of plausible design options and choices between them. There is bound to be a kernel to this work that can't be reduced to procedures or even principles; you have to use your own talents. Still, as a generalisation, the various design activities described in Briefings 4-8 ought normally to be done in a way that exposes and explores *at least some* options, rather than treated as stages on one relentlessly methodical march. But how many major options you go into and how deeply are situation-specific questions.

10. Implementation Approach Decisions

ISSUES

Once you have decided on a sound new design for a process, how do you implement it throughout the organisation? There are two main complexes of issues, distinct but related.

Phasing

Everybody seems to agree that it is usually best to start with a pilot phase — handling actual live cases rather than conducting an experiment — and to introduce the new process throughout the organisation in further phases. This is plausible enough, but it still leaves room for weighty choices.

In some situations there may be one obvious phasing plan, but this is far from always so. It may be best to phase in the whole new process in one branch office after another, but an alternative is to introduce a complete-but-basic version in all offices; and then, after this is live throughout the organisation, to introduce the extra, more refined features making up the full design, in one or more subsequent phases. There are several other possible patterns of phasing too.

Such options are relatively easy to sketch out once thinking is pressed in that direction, and choice between them could be a decision with far-reaching consequences for success or failure. Naturally, the most worthwhile options and the factors affecting decision will vary from project to project.

Change Philosophy

There is a broader issue. The phasing possibilities just mentioned are ways of organising that stretch of the project between the moment that the new process is first seriously used through to the moment when it is safely established in full throughout the organisation. But now extend the horizon to contain the whole period that the process will be used in the rough and tumble of changing business circumstances. Can the process still be modified, and, if so, how? This is to wade into vaguer questions of what can be called change philosophy.

The debating ground is best delineated by an extreme choice. Should you install the re-engineered process throughout the organisation, use it unaltered for a number of years, and at some point in the future, when circumstances are very different, carry out a new re-engineering project all over again? Or should you install the process, use it, but, almost from the start, make changes and improvements, continue to do so, adapting to circumstances, and thus avoid ever having to re-engineer it again?

That there is a difference may readily be granted, but what does it matter? It may well affect decisions of implementation approach. For example, the process introduced throughout the organisation should certainly be sound and workable, but should it perhaps leave room for workers' initiative and taste in certain areas, or include certain variant options? This, if feasible, would fit in with a continuous-improvement philosophy. It may have the change-management advantage of allowing the users of the process some influence on the fine detail of the work they will be doing over the years ahead. On the other hand, it may complicate organisation and co-ordination, and run the risk of anarchy. The tradeoffs will vary from one project to another.

REPRESENTATIVE IDEAS

This section reviews pilots and phasing, topics treated quite superficially even in most full-length books on re-engineering.

Phasing: Typical Views

The natural approach is to begin with a pilot in one part of the organisation; then, once that has gone satisfactorily, to introduce the process in the rest of the organisation — either straight away, or in different parts of the organisation in several phases.

One associated concept is that the design team should hand over to a separate team responsible for organising the pilot and subsequent implementation of the new process. The Bell Atlantic example illustrates this approach. A notable feature is that the implementation team feeds back demands and ideas for further changes in the process to the design team. Hammer and Champy say that whether responsibilities are split between two teams is not important, but that seems implausible. Surely the approach may be good or bad depending on the situation.

Cross et al seem to assume that there should normally be a separate pilot team, and they describe an advanced version of the approach: first, have a pilot in one part of the organisation run by a pilot team; then split this pilot team into two teams, fill up each of these teams with new members and implement the process in two more parts of the organisation; then split these teams to make four . . . and so on. This could be sensible in some circumstances, but uncritical praise for it seems naive. It is dependent on the assumption that no feedback of any consequence will occur. Should the people in one out of (say) four parallel teams wish to send back intelligent ideas for doing things even slightly differently (eg altering the design of a certain input screen to include one extra data item), the complications of change control (evaluating the feedback and propagating the process change to all other teams) could be formidable.

Andrews and Stalick suggest a variant concept: the 'pilot test'. This is, in part, a pilot in the sense understood in the rest of this book; ie running a newly designed process believed to be sound, handling real cases for (say) six months in one part of the organisation — but it is also a test, in the sense that detailed quantitative measurements are made, to see whether the new process lives up to expectations or not. This seems unconvincing:

if it is truly a test, then there must be a fair possibility that performance will be inadequate and the process will have to be withdrawn; but if that uncertainty does exist, less disruptive approaches are preferable: simulation, perhaps, or extensive prototyping with representative but not live cases.

Possible Phasing Patterns

Davenport and Stoddard, in the article about the myths of re-engineering, are among the few to point out that there are at least two dimensions to the choice of phasing pattern. You can stagger introduction of the whole newly designed process to different parts of the organisation (as most writers suggest), or stagger introduction of different portions of the whole process.

For example, suppose the newly designed process falls naturally into three segments — blue, green and red — and the organisation has a number of branch offices. Then there can be a pilot of the blue segment only in one office; the corresponding segment of the old process is replaced, while the remainder continues unchanged. Once this has been done throughout the organisation, the green segment of the process, and later the red segment, can be switched in too.

However, in a comprehensively redesigned new process, the magic-square factor may make it impossible to slot in a new segment while staying with the old versions of the other segments. Recognising this difficulty, Davenport and Stoddard sketch out quite a radical conclusion. Rather than first completing the design and then considering the phasing of implementation, you should *design* the process in such a way that it *can* be implemented in segments. On this view, just as designers of physical products, such as videorecorders or tractors, have to adopt the constraint that it must be possible to manufacture whatever is designed, so designers of new business processes should make only designs that lend themselves to easily phased implementation.

DISCUSSION

Davenport is much the most interesting author of those cited on these issues. Even then, the views in his book are unremarkable, and the interest lies in the later article. His points are well worth considering, provided they are taken as sketches to be worked out, rather than prescriptions for what should normally be done.

Aspects of Phasing Plans

The phasing pattern usually described is that where the whole new process is introduced in one part of the organisation after the other. The preceding notes hint that the possible options are more complex than this; here are some generic possibilities:

● **Business-unit.** The whole new process is introduced in one organisational unit after another; eg successive branch offices.

● **Business-type.** The new process is introduced throughout the organisation, in phases with respect to one type of business after another; eg in the first phase, the process is used only for certain types of customer; in the second phase for certain other types too; and so on.

● **Process-segment.** Segments of the new process are introduced throughout the organisation, in successive phases. First, one segment of the new process replaces the corresponding segment of the old, while the rest of the old process remains in operation; then another segment is switched over; and so on.

● **Process-layer.** 'Layers' of the new process are introduced throughout the organisation, in successive phases. First, a complete new process, but only with basic facilities, replaces the entire old process; then a second layer adds more sophisticated features; and so on.

The diagram illustrates these possibilities impressionistically. A couple of qualifications are in order. First, the phasing pattern need not always be totally pure. If phasing primarily by business-type, you may also stagger to some extent the introduc-

Possibilities for Implementation Phasing

1. Whole process - all types of business - phasing by business-unit

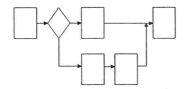

2. Whole process - all business-units - phasing by types of business

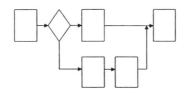

3. All types of business - all business-units - phasing by process segment

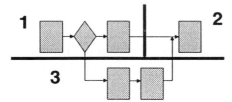

4. All types of business - all business-units - phasing by process layer

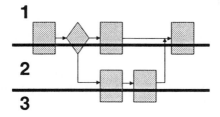

tion of the process at different business-units. The possibilities given are really four *aspects* that help in clarifying feasible phasing options. That said, for incisive decisions of implementation approach, it is usually best to make an explicit decision that one of these four aspects should be the one that dominates detailed planning.

Second, the analysis omits another complication. You may in some cases, with due circumspection, occasionally complicate the phasing plan by allowing temporary versions of part of a process. Suppose phasing is essentially by process-layer: the first phase can use a temporary layer 1; later, instead of sliding in a layer 2 to complement layer 1, you discard layer 1, and replace it by a new version of the process, that contains all the layer-1 facilities, though different in detail, together with additional layer-2 facilities.

As this suggests, there may be surprisingly many phasing options to choose between. In reaching a decision for any particular project there are numerous factors to take into account:

• In some situations the task of making a good process design may be so exacting that design in layers or segments for phased implementation, as suggested by Davenport and Stoddard, may be asking too much. It could be quite rational to design first and only consider the pattern of phasing afterwards.

• For a truly ingenious process with strong magic-square properties, where all the parts are closely interlocked, many of the generic options suggested above may be just impossible.

• Phasing based on process-layer is generally the most demanding to plan and arrange, but may have strong counter-advantages. An advanced process design with numerous extra, innovative features may entail substantial investments in new hardware and software with long lead-times; if so, phasing by process-layer may be a very strong option.

The Problem of the Development Gap

Books often give the impression that almost as soon as the design of the new process is settled, it can be implemented in the organisation — in a pilot phase no doubt, but nevertheless handling live cases that, for the health of the business, must be processed perfectly. But many re-engineered processes depend on new IT arrangements, above all new software. Often — if the redesign is truly radical — the software's lead-time is many months. This brings the unpleasant prospect that perhaps a year may pass between the end of design and the start of implementation, while analysts and programmers toil, more or less invisibly to the rest of the organisation, to get software ready and bug-free. This gap is a ghastly thing from the organisation-wide, change-management point of view; enthusiasm is bound to cool, scepticism to mount.

What can be done? First, at least recognise the problem, instead of ignoring it as many writers do. Second, make it a conscious decision-factor in comparison of process design options: if design A is just a bit more radical and innovative and financially beneficial than design B, but A will take a year to develop and B only six months, then perhaps B is the better buy.

Third, take account of the development gap in deciding phasing plans. By juggling combinations of the four aspects given above, it may be possible to sketch out a dozen plausible options for phasing in an elaborate process in a large organisation. Those options that stress the process-layer or process-segment aspects will, in general, have shorter development gaps than those stressing either of the other two aspects. This consideration could be decisive in knocking out many otherwise attractive options. In fact, if substantial development work is unavoidable, there may be little choice but to go for whichever phasing option brings the shortest development gap.

'Quick Strike': Precept and Tradeoffs

Some consultants are partial to the 'quick strike' (aka 'quick kill' etc) precept: plan implementation of a new process so that, quite early on, some small but worthwhile, unquestionable, demonstrable benefit is achieved. That will generate positive feelings about the whole implementation, even if it takes several years to be fully realised.

For example, suppose you are implementing a completely new process for taking and scheduling adverts in the magazines of a publishing group. One defect of the old process is that, when a customer wants to advertise for the first time in a certain magazine, it can be a laborious task to check whether that customer has ever advertised before in one of the group's other magazines. (This is important because there could be an out-standing debt; in any case, it is undesirable to hold two accounts for the same customer.) As top priority, you introduce a facility — forming only a small part of the entire redesigned process — which alleviates the drudgery of that checking. People are delighted with this quick-strike improvement, and some sceptics in the organisation concede that perhaps change is not such a bad thing after all. Drawing on this emotional capital, you can phase in the rest of the process more methodically, even sedately, in a more positive atmosphere.

Yes, of course. How could anyone argue with that? If the precept is left in the superficial form just given, then there is nothing very clever about it. The interest comes when it is examined more closely. Suppose there is a choice between two possible phasing plans, and they are equal in respects such as long-term return-on-investment, risks, disruptions, and so on — but plan A brings a quick strike, similar to the example above, and plan B doesn't. Then, obviously, A is the choice. But suppose plan A with the quick strike is assessed as — in the long run — somewhat more complicated and messy and costly than the methodical B. What then? This is a tradeoff situation. To decide you have to strike a balance between factors that are incommen-surable. The positive benefits generated by A, if successful, have to be set against the safer virtues of B. The decision may go either way, depending on the circumstances.

A consultant who talks knowingly about quick strikes but misses this tradeoff is just purveying a stale cliché. It is as absurd to say that every project must go for a quick strike, whatever the situation, as it is to say that phasing should always be decided on the basis of hard factors alone with no attention at all to soft ones. But a consultant who helps an organisation understand the tradeoff factors, and assess them in the particular case may well be worth using.

This suggests that the quick-strike precept is essentially a change-management, rather than directly-rational, one. It can affect your phasing decision if, and only if, you allow the possibility that softer factors, such as the building up of credibility, enthusiasm, commitment and so on, may conceivably outweigh harder factors such as cost and timescale. Like most change-management precepts the quick strike is only one element in the decision-making calculus.

The quick strike is only a motivation, not, in itself, a phasing plan. In one situation, the best means of achieving the soft quick-strike benefits may be to introduce the whole new process in one branch office out of the hundred, so that it can be a shining example to the rest; in another, to introduce one segment of the new process in one office; in another, to introduce a small segment of the process throughout the organisation; in another, to introduce a temporary, partial process, that will be thrown away when the full process is implemented.

Kinds of Process, Phasing by Business-unit or Business-type

Decisions on phasing pattern depend on the situation, and situations vary considerably, but is it perhaps possible to assess how certain kinds of situation tend to call for certain kinds of phasing? Any such generalisations must be subject to the disclaimer that, in any specific case, things may be entirely different, but still, analysis by the four kinds of process is worthwhile.

One question to ask about any process is this: How feasible are the business-unit or business-type patterns of phasing?
● With a **transaction-based process**, both patterns (particularly business-unit) are often feasible and natural. That is why this is the approach usually mentioned (or assumed) in writings about re-engineering.

● With a **matter-based process**, these patterns of phasing are often attractive too — but not always. If the process re-engineered is used by just one town hall or by one head-office department, phasing by business-unit is much less of an option. Also, phasing by business-type may be undesirably messy; eg if staff in a bailiff's office have to remember to treat their bad-debt matters in different ways, according to whether they are being handled by the old or new versions of the process. It is true that for a process such as the development of OLS products at AT&T, with separate design teams, phasing can be by business-unit. But difficulties may still arise if all teams share or depend on some common resource, such as the inventory or production system; there may be an unacceptable degradation of service for teams still using the old process, or a misleading, untypical environment for those with the new.

● With a **project-based process**, such as developing a new aeroplane or negotiating a mining concession with a government, the cases are few but weighty, and phasing by business-unit or by business-type are not options. Put another way, it is fairly obvious that you will start off using the new process design with just one case, and, if all goes well, probably use it for the next case that comes along, perhaps several months later.

● A **facility-based process** tends to supply information, that is either valuable or not, as opposed to handling live cases either correctly or not. Thus, once the facility is considered reliable and useful, there is no strong reason for any elaborate phasing by business-unit or business-type. The main constraint on speedy organisation-wide use may just be the effort of getting round and training people to use the facility.

Kinds of Process, Phasing by Process-layer or Process-segment

Another question about a process is: How feasible for this process is phasing by process-layer or by process-segment?

● With a **transaction-based process** the possible arguments against this approach are: that the process has magic-square characteristics, making partial implementation awkward or impossible; that the demonstrable payoff only comes when most or all of the new process is running; that many people would rather get a major change over in one heave, than go through prolonged turmoil in several phases. These arguments can apply to some degree with any kind of process, but are, on the whole, likely to be strongest with a transaction-based process.

• A **matter-based process** handles cases that are inherently more complex. This tends to reduce the force of the arguments just given. There is a greater chance of starting off with a productive new process, superior to the old but still with only basic facilities, and later implement extra *de luxe* features (eg expert-system based intelligence to improve certain crucial decision-points; or extra telecoms facilities, linking up previously separate people). Again, many of the people using the process may have more demanding and responsible roles than in a transaction-based process; if generally sympathetic to the change, they may well grasp and welcome the logic of implementation in process-layer or process-segment phases.

• Many **project-based processes**, particularly those for product development, where data about parts of a machine must be tied together thoroughly and rigorously, have strong magic-square properties — a powerful argument against phasing by process-layer or process-segment. Other processes, where integration of information can be less rigorous (eg developing an advertising campaign) typically need a good deal of prototyping before implementation proper. Here rigorous integration of information is a necessary but only minor objective; the real challenge is to tune the process so that people can access information easily and flexibly. Again the natural course, once you believe you have a sound new process, is just to introduce it without any phasing by process-layer or process-segment.

• With a **facility-based process** some of the main distinctions — between prototype and pilot, and between a phase-1 and a phase-2 version of a process — become far weaker. A facility like IBM's information system about Latin America needs, as a minimum requirement, to be solid and reliable, and reasonably convenient and helpful. Once those criteria are met, the whole system can be introduced without much phasing — but with the proviso that people will very likely discover new facilities and information that they want, or will suggest new modes of use. Thus further developments will probably occur, but not necessarily in neat phases.

REPRESENTATIVE IDEAS

This section examines how decisions of implementation approach can be affected by broad considerations of change philosophy.

Some Concepts of Change Philosophy

Should re-engineering be seen as a once-and-for-all event or just one part of an endless stream of change? This rather vague question is associated with the notion of continuous improvement:

● If, after the re-engineered process goes live, you continue to monitor it intensively and make incremental changes to it, so that it keeps abreast of changes in business conditions, technology possibilities, people's requirements and so on, then there may be no need for re-engineering for a long time, if ever.

● If, on the other hand, you regard re-engineering as a kind of purifying experience, needed regularly to maintain the vitality of the organisation — as some people once regarded war — then there is much less point in expending resources on incremental improvement and monitoring.

Not all recognise this distinction squarely. For example, Cross et al talk a great deal of continuous improvement, through monitoring and measuring, but don't draw the natural conclusion that this should make re-engineering into a near once-and-for-all event.

Hammer and Champy quote somebody at Hallmark '. . . it is a once-in-a-lifetime opportunity. We are building the organizational capability that will enable Hallmark employees to react swiftly and successfully to continuous, unpredictable change.' What could be clearer than that? And yet the authors' comment on the Hallmark example is '. . . re-engineering is not a one-time trip. It is a never ending journey, because the world keeps changing. Processes that have been re-engineered once will some day have to be re-engineered all over again. Re-engineering is not a project; it must be a way of life.' Can re-engineering be a once-in-a-lifetime opportunity, and yet, at the same time, also a never-ending journey? Possibly, but the debate is entering regions so misty as to be impenterable.

Morris and Brandon make the puzzling statement that in future re-engineering will be done on a routine basis, but this follows from their eccentric position that any kind of change, however large or small, is a re-engineering change, provided it is done within their standard methodology.

Design Finish

There is a tendency to see a design team, composed of the best and the brightest people in the organisation, conceiving an ingenious new process, which all the others will accept and use, as designed, in their daily work for years ahead. Nobody advocates that in so many words, and there is much talk of inspiring the workforce and getting them to 'buy in' to the new processes, but often the measures advocated are of purely emotional force — analogous to flags and oratory and stirring music in time of war, but not to democratic consultation about grand strategy, or to debate within a military unit about tactics for capturing a given objective.

Davenport and Stoddard point out that large design questions (war strategy, to pursue the analogy) must inevitably be decided by a small group of the brightest and the most senior, but that there is much to be said for leaving smaller details undesigned, to be filled in, or perhaps determined in a long-term evolutionary process, by those who will have to do the work for ever afterwards. This seems a shrewd concept, but it remains undeveloped.

Without examples its feasibility and implications don't emerge for consideration.

For completeness, Morris and Brandon also seem to hint at the same concept, but their almost example-free book inspires little confidence.

DISCUSSION

In this region change-management and directly-rational factors are strongly intertwined. This section concentrates on the directly-rational, leaving the most general change-management factors to their own briefing.

Comparing Change Philosophies

As the following diagram suggests, there seems to be a choice between the approach of semi-finished design with continuous improvement as one extreme approach, and clean design, limited ongoing improvements, and thus more frequent re-engineering, as the other.

One influence on general decisions of that sort is the magic-square factor. The more firmly the process has a magic-square character, the less appropriate tuning and customising and continuous improvement will be, since apparently harmless minor changes may have major negative repercussions. Conversely, if a process does break down very cleanly into segments, then allowing people freedom to tune discrete portions may be attractive.

In principle, any process may need to evolve, because it is always possible for some outside factors to change. For instance, it is true of most processes that a government might conceivably alter taxation in some way that affected the handling of cases. But certain processes may have a clear need to support change that goes far beyond this general consideration. A process assisting product development, with (say) five new products a year, may be fine for the first product; then need substantially changed facilities for the second; some but fewer changed facilities for the

Re-engineering and Continuous Improvement

At one extreme:

Process as designed
by project team,
implemented, and used
unchanged for four years

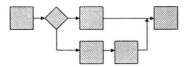

Re-engineered process,
introduced four years later

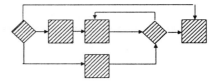

At the other extreme:

Process as designed
by project team, and
implemented for live use

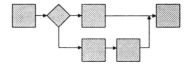

Process after
six-months pilot use

Process as evolved,
one year later

and so on . . .

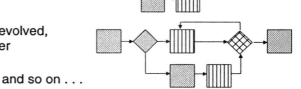

third . . . By the fifth product, it may be apparent that all these varied facilities can be rationalised as a range of optional features, so that for any subsequent case it is only necessary to configure the system with the desired features.

Often the softer change-management factors may be the more telling influence on change philosophy. The option of semi-finished design (as opposed to one unalterable design that is handed down) seems more likely to encourage workers to 'buy in' to the new process. Even if this is true in general, much depends on the situation. Some workers may be irritated at being given a semi-finished, multi-choice process, and may prefer to know exactly what they are supposed to do. Others may tune the process in calamitous ways that undermine confidence.

It may actually be misguided to decide firmly for one philosophy or another. You can make some forecast of the magnitude of changes likely to be caused by uncontrollable, outside factors, but you can easily be wrong. However, one thing you can control is your own rhetoric. In one situation, it may seem psychologically advantageous to raise the stakes by stressing that this re-engineering project is the once-in-a-lifetime chance to dominate the industry, that it can succeed if everyone supports it, but that, if not, the business is condemned to mediocrity or worse. In another situation, it may seem better to introduce the new process more modestly, saying that, though better than the old one, it probably won't be ideal, but there will be plenty of scope for tuning it over the years ahead. The point is that the choice between such attitudes need not have much basis in facts or logic; it can be made primarily on psychological grounds.

Kinds of Process, Change Philosophy

Even so, reason can still make some contribution to decisions of change philosophy. In particular, the kind of process makes a difference:

● With a **transaction-based process**, it is relatively plausible that, because of the magic-square factor, a jump to a radically new version of the process may be desirable from time to time, to keep up with business and technology changes. That this kind of process dominates re-engineering literature explains the bias in the rhetoric towards abrupt change.

● A **matter-based process** deals with more complicated cases, that are often more dependent on human discretion and judgement. This tends to favour the two notions of continuous improvement over time, and the granting of flexibility to users about how exactly they do things. Frequent re-engineering of such a process seems unlikely to be appropriate.

• In a **project-based process**, for any given case (ie project) you will probably use one version of the process and stick to it, without much continuous improvement to the process during the project. Why? In a process to develop a technology-based product, such as an aeroplane, the requirements of strictly integrated information make it dangerous to meddle with the process in mid-project. At an advertising agency, the same may apply for different reasons: many likely requests for improvements will be technically demanding (eg more sophisticated facilities to merge several high-resolution graphic images in one advert) or need experiment and practice (eg better statistical procedures to correlate types of advert to social group), and are preferably not handled in mid-project.

However, projects being both complicated and varied, you may very well make an amended version of a project-based process, from one project to another. You may even maintain different versions of the same process, so that, for any new project, the most suitable can be chosen.

• A **facility-based process** is inherently a good candidate for continuous improvement. It is free of the strong constraint of having to process specific cases correctly, and usually deals with vague things such as information and knowledge. Desiderata such as relevance, flexibility and convenience are dominant. Any alert body of users will find endless opportunities for incremental improvement.

CONNECTIONS

For the difference between pilots and prototypes see Briefing 4.

Decisions of scope, design approach, process design and implementation approach may all be influenced by both directly-rational factors and by change-management factors. The latter are probably more influential in decisions of implementation approach than in decisions of the other kinds. See Briefing 11.

11. Change Management Factors

The term 'change management' is used in a variety of ways. If meant to cover anything whatsoever to do with change, then it embraces this entire book and more besides. The term gets a much sharper edge when it denotes just those factors in the management of change that are concerned with human feelings. That is the topic of this briefing.

Advice about Change Management

There is no doubt that poor handling of change-management factors can ruin a re-engineering project. Plenty of advice about change management can be found in books and articles, but there is one great difficulty: change-management factors are not readily susceptible of dissection on paper or in debate.

Everyone agrees that top-management commitment is vital. Also, you should watch out for the danger that corporate culture and management attitudes may stifle re-engineering from the start. Then again, it has to be accepted that radical re-engineering usually makes at least a few people unhappy. But aren't these and dozens of similar points mere platitudes that will already be obvious to anyone competent to re-engineer a process?

If it has never occurred to you that top-management commitment is necessary to success in revolutionising a huge chunk of the company's business, then you are probably inadequate in many other ways to play any leading part in the project. Reading a book or attending a seminar that points out the obvious is hardly likely to save you; you would do better to resign in favour of somebody else with a better grasp of reality. And if you do already know that top-management commitment is important, what more remains to be said and discussed? This doesn't seem to be a topic where any intriguing contests between plausible, opposing doctrines can occur.

Nevertheless, openings for stimulating discussion can be found. First, some apparently obvious tenets of change management turn out, on critical examination, to conceal seething tradeoffs between competing desiderata: yes, top-management commitment is desirable, up to a point; but democratic consultation and consideration for people's feelings are also worth having, and may at times act as contrary forces. Thus critical thinking about apparently obvious advice can yield insights that affect decision-making.

Radical and Less Radical Change

A sharp contrast is often drawn between radical re-engineering and incremental improvement. If valid, this distinction seems likely to be important. Davenport's influential book, for example, starts off with a table comparing the two types of change, and constantly refers back to the distinction. But is it a useful distinction?

This briefing will argue that there is really no chalk-and-cheese difference here. If that is true, writers, consultants and managers who rely on the distinction may be misleading their audiences, and basing decisions on unreliable foundations.

But not necessarily. Even if a sharp distinction between radical re-engineering and incremental improvement could be discredited as a rational concept, there might still be some purpose in talking as if it were valid. In war a government may call on people to contribute tin or brass objects to be melted down for armaments — not because it is a cost-effective measure, but because it strengthens civilian morale. Similarly, some rhetoric about radical re-engineering may be essentially nonsense, but still effective in influencing some people: encouraging them to think inventively, challenge preconceptions, feel enthusiastic, and so on. In short, the rhetoric may be a change-management factor.

Thus any debate about the distinction between radical re-engineering and incremental improvement needs to argue on two levels: first, whether the concept is sound or not, at the rational level; and second, what effect it may have as rhetoric that influences human feelings. This is one good example of the peculiar role of change-management factors in the decision-making calculus.

REPRESENTATIVE IDEAS

The question seems quite stark: is most of the advice given about human factors in re-engineering mere platitude, or is it not?

Representative Advice

Cross et al set out a scale of 18 gradations of possible behaviour towards change that may be encountered within an organisation — ranging from enthusiasm at one extreme, via working to rule, through to deliberate sabotage at the other extreme. The intention is presumably that if you know about this scale you will be better equipped to manage change than if you don't. But how? Any adult is already aware that the kind of reactions described on the scale can occur in human behaviour. What is gained by arranging them as 18 gradations? The authors don't say.

Andrews and Stalick devote seven pages to 'terrorists and saboteurs' of re-engineering projects. They describe five species of nefarious behaviour: lone ranger, game player, opportunist, technocrat and pretender. For each they suggest 'defence strategies': eg look the villain in the eye, and threaten to go to his or her boss if the terrorism doesn't stop. This might be fruitful if the five categories were generic ones; but all the authors have actually done is describe five particular, awkward individuals they happen to have met. There is no general relevance.

Change Management: Insights and Platitudes

Communicate to people at every level in the company.

Communicate truthfully and fully.

Create an environment that encourages employees to come up with innovative ideas.

Don't assign someone who doesn't understand re-engineering to lead the effort.

Don't bury re-engineering in the middle of the corporate agenda.

Don't dissipate energy across a great many re-engineering projects.

Don't lose momentum.

Don't neglect people's values and beliefs.

Don't skimp on the resources devoted to re-engineering.

Ensure all employees participate in a mindset change.

Ensure project team members have healthy interpersonal relationships with each other.

Ensure project team members have healthy interpersonal relationships with those outside the team.

Ensure senior management commitment.

Expect resistance, and regard it as a sign that people accept that you are doing something significant.

Fine-tune project team communication skills.

Force people to confront their biases and assumptions.

Foster commitment and ownership at all levels.

Get a good mix of people in the team — in terms of knowledge, skills and personality.

Grasp people's feelings and sense what is politically or psychologically feasible or infeasible.

Have some outsiders and mavericks to ask naive and outrageous questions.

Involve those people regarded as the best and brightest, as a clear signal that top management is serious.

Listen and ask questions, rather than telling everyone what they should do.

Recognise that some people will fear change.

Treat people with dignity.

View complaints as appeals for information and support.

The table presents a selection of imperatives (ordered alphabetically) drawn from various sources where perhaps a similar line of scepticism may prevail — or perhaps not. For many more precepts in similar style see the book by Andrews and Stalick.

A minority of published material seems to offer rather more to think about than the above. Davenport and Short point out that re-engineering is like any other type of change in the organisation, and thus many straightforward, general pieces of advice apply to it. But most other changes, even large ones, occur within one clearly defined business unit. With re-engineering, on the other hand, the stress on cross-organisational process increases the number and variety of stake-holders and thus the complexity of the human factors. On this argument, a re-engineering project will very likely pose greater challenges than another change project of the same size — as measured by financial investment, number of project activities, number of people involved etc.

The article by Earl and Khan, and the later one by Davenport point out an intriguing paradox. On the one hand, some of the organisation's IT people may be the best qualified to lead a re-engineering project; but on the other, their dominance may be psychologically or politically inappropriate. For instance, the IT department as a whole may have a reputation (deserved or not) for being too bossy or too remote or giving too little weight to non-technical factors. Therefore — though it may seem strange — you should not necessarily put somebody from IT in charge, even if that person is really the best qualified.

DISCUSSION

There is no point in this section reviewing the ins and outs of all the change management points, platitudinous or not, presented above. Instead it offers a kind of primer with examples on critical thinking about this kind of discourse.

Human Factors: Two Examples

Cavanaugh's article describes a power utility that set out on an eight-year 'Journey to Excellence', giving each year a Chinese-style name; eg 1990 was called 'The Year of Awareness'. Hammer and Champy praise Taco Bell for enunciating the corporate vision: 'We want to be number one in share of stomach.' They aver that every company engaged in re-engineering should think up something of that kind.

Scepticism is in order, but not outright dismissal. There may well be a few organisations where Taco Bell's slogan would not be regarded as puerile and naff. But nobody, not even Hammer and Champy, could believe that slogans and year-names actually advanced *thinking* about process design, or supported a chain of reasoning, or opened up new options for consideration. A slogan affects people's *emotions*: it may inspire them, but it may just as well obliterate respect for those who promulgate it. Either way, its force is emotional, not rational.

This point about slogans and year-names has general relevance. The key to appraising the items of change-management advice in the previous section is to notice that most of them are about affecting people's emotions rather than their intellects.

Davenport's book goes as far as suggesting that the entire change effort should be viewed as a public relations campaign — akin to selling a product or a political candidate. Most people would agree that doing either of the latter usually does require some degree of cynical manipulation: selective presentation of truth (or even half-truth), appeals through carefully crafted slogans, restraint from open, honest argumentation, and so on.

It probably is correct to say that selling the change effort should be viewed in that way — to some degree. If everybody is entirely honest and open about every single thing all the time, and nobody ever stoops to communicating in ways that appeal to an audience's emotions, then the change initiative may well fail. The interesting issue for decision is *how far* to go in this direction. If you manipulate people too cynically, they may find you out in the long run and confound all your schemes.

As Earl and Khan say, an organisation's IT managers and analysts may, from an objective point of view, have a good claim to lead re-engineering. They may well be more shrewdly analytical, experienced at managing projects, knowledgeable about technology factors, and objective about boundary disputes between departments than most other managers. But, from a practical point of view, it may be better if they are less dominant than their qualities merit. It may in many instances be advisable to give outward signs that this is not an IT project of the kind the organisation has known in the past, but one that is far more business-oriented.

This is a subtler point than the one about slogans. The argument is not that the IT people may be *inferior* at leading a team of bright people to design neat new processes (they may be, but then no delicate point of human factors arises at all). Rather, the argument is that, even if the IS people are *better* qualified, you should perhaps pick some less-good, non-IT leader, as a symbolic gesture to the organisation as a whole. Of course, it may be expected, the nominally subordinate IT people may still become beneficially dominant within the confines of the brainstorming chamber. This may seem an odd or cynical attitude, but something of the sort is found in many human organisations. For instance, in the Imperial German army an aristocratic commander (eg von Hindenburg) often had a chief of staff selected on merit (eg Ludendorff), with more effective power.

From Platitude to Tradeoffs: Examples

As a thought-experiment suppose the issue for decision is the makeup of the team to do the design work — a decision of design approach. You provisionally choose a team of seven people on directly-rational grounds; this is the best set of people for the task, bearing in mind such attributes as intelligence, creativity, energy, knowledge of the business etc. The question is: Could change-management factors cause you to alter your directly-rational choice of team members?

At one point in his book Davenport advises that the selection of a team should be done in a way that maximises acceptance of change by the rest of the organisation. Taken literally this means that you should consider no other factor other than maximising acceptance of change: candidates' skills and knowledge for *designing* processes don't come into it, except in so far as those qualities help persuade others to accept a new process. Somebody who is poor at being sympathetic and polite with other people — even though outstandingly good at inventing ingenious processes — has no chance whatsoever of being selected.

This seems too extreme a view to be acceptable. It provokes the retort that you should take account of *both* the design-skill and the maximise-acceptance desiderata rather than just one or other of them. But this less extreme positiion does allow that, yes, change-management factors might cause you to alter your original choice of team.

To pursue this line of inquiry take one of the items from the list given earlier: 'Involve those people regarded as the best and brightest, as a clear signal that top management is serious.' Is it an empty platitude or has it some decision-supporting force? If taken to mean that, other things being equal, a bright person is more use than a dim one, then it is inconsequential. But the real interest is in the words 'regarded' and 'signal'. Also 'best and brightest' can have nuances beyond its literal meaning; Halberstam's famous book of that title puts the thesis that it was the young, impressive, articulate, stylish, pushy, well-educated, upper-class men that led the way into the Vietnam quagmire. In other words, if that thesis is correct, the most ostentatiously clever men were actually very foolish. Thus *being regarded* as clever need not be the same as *really being* clever. But, the argument can run, if the intention is to signal top-management commitment then the perception will be more important than the reality.

Recognising this, you might discard Anton — a 45-year old branch manager, shrewd, experienced, awkward, phlegmatic, unpretentious — from your provisional team, and instead choose Bernard — manifestly one of the best and brightest, aged 30, impressively articulate, well up in management theory, certainly on the fast track, in a head-office staff function — even though, if truth be told, Anton is likely to contribute rather more to an effective new design.

The point is that there *can* be a conflict between the choice of the best team *per se* for designing a new process, and the one that sets the right tone and style for a radical, business-transforming, re-engineering project. Resolving the tradeoffs here can call for fine judgement. (Moral questions about the possible injustice of not putting the most deserving people in the team are beyond the scope of this book.)

Further Platitude-tradeoff Explorations

There is another twist to the 'best and brightest' theme. Taken in the Halberstam sense, it will tend to conflict with other accepted tenets, such as 'foster commitment and ownership at all levels', 'ensure all employees participate in a mindset change', 'communicate to people at every level in the company' and perhaps even 'treat people with dignity.'

These too have some obvious merit, but they are tradeoff factors not absolutes. Despite the gushing idealism of some management texts, the most effective way of designing an ingenious new process is *not* necessarily for hundreds of employees of all levels of ability and experience to spend many hours in debate. Often the efficiency of having an elite group make inventive process designs has to be weighed against a conflicting factor: the desirability of involving a wide circle of people in the new way of doing things.

For a further complicating twist, take these two items: 'Have some outsiders and mavericks to ask naive and outrageous questions'; and 'Force people to confront their biases and assumptions.' The trouble is that, whatever some zealots may say, it is possible to go too far with naive and outrageous questioning, and bias-confrontation. If you aren't careful you may flout other tenets, which seem equally plausible: 'Grasp people's feelings and sense what is politically or psychologically feasible or infeasible'; and 'Don't neglect people's values and beliefs.'

The force of these latter considerations is easily seen in any debate about health service reforms in almost any country. Many options and factors that occur to the perceptive outsider are never raised, or if raised, assailed so bitterly that proper debate becomes impossible. So it can easily be with a re-engineering project. Encourage iconoclastic behaviour too much, and there may be turmoil, suspicion, resentment and even sabotage. But eradicate maverick influence altogether, and many ingenious design possibilities may never be noticed. The right balance will vary from project to project.

Thus the obvious-seeming tenets listed earlier conceal a seething mass of tradeoffs, whose untangling might take a whole book in itself. To sum up:

● Establishing an elite 'best and brightest' team (in the Halberstam sense) can be a valuable, palpable sign of top-management commitment.

● But a second factor, fostering wide commitment and ownership, comes close to being intrinsically contrary.

● And you should of course bring a third factor into the tradeoff calculus: the simple desirability of having the people in the team who are most likely to conceive the best new designs most efficiently.

● Fourthly, potentially in friction with any of the above, you need to stimulate iconoclasm.

● But without doing too much violence to a fifth factor: navigating around people's unchangeable prejudices (though only to a certain extent, of course).

▼ The point of the above discussion is to suggest that thinking critically about sweeping generalisation and apparent near-platitudes isn't just a negative activity. Done properly, it will very likely expose some crucial tradeoffs. This applies to other change-management topics besides those discussed, and to many other areas of management too. ▲

Charting out Top-down and Bottom-up Issues and Approaches

Much of the above can be distilled into the general points that top-management commitment is necessary, but so is the commitment of ordinary workers. These two things are not necessarily in conflict, but when the terms 'top-down' and 'bottom-up' creep in, the discussion is often coarsened into confusion. As the following diagram suggests, these terms may be used in debating at least three different subjects:

• **Management concern.** Here 'top-down' means that the most senior managers in the business initiate the re-engineering and drive it through. 'Bottom-up' means that the stimulus and energy come mainly from managers other than the most senior (though nobody believes that it is possible to do without *some* top-management support).

• **Design approach.** Here 'top-down' means that a relatively small design team firms up the detail of the process before it is given to the rest of the organisation. 'Bottom-up' means that many people beyond the design-team are involved in design activities, eg by trying out prototypes, while significant options still exist. It may also mean that some details of the process are left open for workers to deal with in ways they choose, even when the new process is live.

• **Scope.** Here 'top-down' means that there is a master plan co-ordinating a number of re-engineering projects; the alternative (for which 'bottom-up' is not a good term) is for each re-engineering project to be linked only loosely, if at all, with activities elsewhere in the organisation.

These three are separate variables that need not hang together: there could be top-down *management support* for a bottom-up *method of design*; or projects with a top-down *method of design* and bottom-up *scope*; and so on. Failure to disentangle these things can lead to confusion. For instance, the article by Caron et al on the CIGNA insurance company claims to describe successful re-engineering that is less top-down than Hammer and Champy recommend; but when the three variables just given are

Top-down and Bottom-up

There are three separate aspects:

MANAGEMENT CONCERN

Top-down
Senior managers
driving everybody
else to
re-engineer

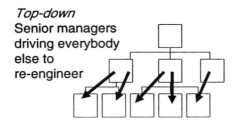

Bottom-up
Middle and junior
managers urging
seniors to be
more radical

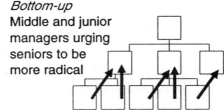

DESIGN APPROACH

Top-down

Small Team

designs in detail

New Process

which is handed down to

Rest of the
Organisation

Bottom-up

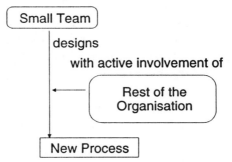

Small Team

designs

with active involvement of

Rest of the
Organisation

New Process

SCOPE

Top-down

1. Make a master-plan design.
2. Develop detail of each
process individually.
3. Fit them all together.

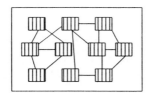

Bottom-up

1. Make a rough process map.
2. Pick a process.
3. Develop a new design for it.

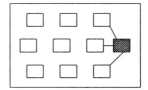

taken separately the authors' claim is far from proven. The main points from that article are these:

- **Management concern.** CIGNA has a variant that is more like sideways-across than top-down or bottom-up: a group of re-engineering experts, with great prestige and momentum, sweeps through the organisation from project to project.
- **Design approach.** Much stress is laid on the involvement of ordinary workers during the design work — with prototypes, or with comments on post-it notes attached to model diagrams. This is bottom-up.
- **Scope.** In one of the company's main business-units everything was re-engineered. There was a top-down master plan, on which the first six areas for implementation were identified. The article is frustratingly silent about the nature of the master plan (three pages or 300?) and about the means of co-ordinating the design work in the six areas.

As this suggests, on a great many re-engineering issues for decision there can be no unequivocal general advice. The challenge is to recognise options — very often roughly describable as 'top-down' and 'bottom-up' choices — and their tradeoffs, and decide the balance of advantage in the specific circumstances. That some factors of influence are soft, change-management ones, adds zest to the decision-making.

REPRESENTATIVE IDEAS

There is much talk about the difference between radical re-engineering and the much-less-radical incremental improvement. Davenport begins by setting out this antithesis, and it recurs throughout his book. Hammer and Champy have some powerful things to say about it. Most writers on re-engineering raise it.

Rhetoric: 'Obliteration' and 'Leninism'

Hammer's original, influential article exhorts everyone to obliterate outdated practices, break away from conventional wisdom, welcome fanaticism, and more in that vein. Strassmann's trenchant article, 'The Hocus-Pocus of Reengineering', by contrast, advocates evolution rather than revolution, and likens re-engineering's radical promoters (presumably including Hammer) to Robespierre, Lenin, Mao and similar 'political hijackers' whose revolutions did more harm than good. Obeng and Crainer, muse enigmatically: 'Perhaps (writers and consultants) imagine the corporate equivalent of the French Revolution. In fact, re-engineering is likely to be a quiet revolution — akin to pulling down the Berlin Wall.'

By comparing their own work to Adam Smith's and calling it a manifesto for business revolution, Hammer and Champy leave no doubt of their conviction that re-engineering is more important and desirable than any other form of change.

Sometimes Hammer and Champy even want to lay an embargo on any change to a process, unless it be so radical that it count as re-engineering. They hold that a 10% improvement, even at practically no cost, is best avoided. Why? Mainly because such marginal improvement reinforces a culture of incrementalism, which (the argument goes) is bad because it leads to a company with no valour. (The analyst of argumentation will not be satisfied with this reasoning. It begs the question of why a company that has a culture of incrementalism, and is successful, should *need* any valour.)

No authority quite says in so many words that re-engineering should always be as radical as possible, but many imply it. Johansson et al say 'everything must be up for grabs', and keep talking about breaking the china. Hammer and Champy are in favour of 'disregarding all existing structures and procedures and inventing completely new ways of accomplishing work'. Towers says that the full benefits of re-engineering are achieved by 'courageous people . . prepared to question the underlying rules and principles by which businesses have been managed for the best part of two hundred years'.

Moderate Positions

Though more moderate, Davenport's book makes a big point of a contrast between process improvement and re-engineering (which he refers to as process innovation):

● **Process improvement** is incremental; its starting-point is the existing process; it takes a short time; its scope is narrow; it has bottom-up participation in design, ie relatively junior people suggest changes; it has moderate risk.

● **Re-engineering** is radical; its starting-point is the clean slate; it takes a long time; its scope is broad; it has a top-down design approach (since a view from a high vantage point is needed to visualise radical changes); it has high risk.

This antithesis is made time and again. The book points out consistently that both types of change are necessary, that trying to re-engineer everything is usually foolish, and that the choice for any given process should depend on the circumstances.

Occasionally the author seems to stumble. At one point re-engineering is compared to radical surgery, such as a triple bypass operation, removal of a large malignancy or an organ transplant; this being so, it can only be accomplished when the leaders of an organisation believe that current modes of operation are a threat to the survival of the company. That, if taken seriously, would mean that many businesses will never re-engineer any process (just as many humans go through life without needing radical surgery), and most others will need it only at rare intervals. However, this conclusion seems inconsistent with the rest of the book.

In the later article and interview Davenport again takes the balanced position, but expresses it more vividly and convincingly. He gives examples of moderate changes that can achieve significant benefits, without the trauma and risk of re-engineering, and he attacks Hammer's claims for the intrinsic superiority of re-engineering over incrementalism.

Clemmer's article distinguishes the two types of change, and says that to argue whether one or the other is inherently the more important is like debating whether your right leg or left leg is the more useful, or whether to use only addition or multiplication. On this view, the crucial thing is to decide which to use **according to the situation.**

DISCUSSION

The thinking summarised above rests on the notion that there is a sharp distinction to be made between radical re-engineering and much-less-radical incremental improvement. But is that so?

Comparing Kinds of Change

From the above contributions three incompatible positions can be extracted:

● **Re-engineering** is much more important to the average business than incrementalism — because its effects are much greater, and because most businesses need it to a fair degree. Even if incremental improvements are sometimes necessary, they are far less important. This is the impression given by Hammer and some acolytes.

● No, **incremental improvement** is much the more important to the average business. Re-engineering is risky and has other undesirable side-effects; therefore it is only a last resort, applicable to a small minority of situations. This is the (statistically speaking) unusual position taken by Strassmann.

● No, **both types of change** are important to most businesses, and there is no reason to be prejudiced in favour of one or the other. You should decide each situation on its merits. This is the position of Davenport and of Clemmer.

But all these three are infected by the questionable assumption that a sharp distinction exists between incremental improvement and re-engineering. Davenport's position (and that of the others named and many others too) only makes sense if, as illustrated in the diagram, it is true that any ranking of change initiatives (or perhaps only coherent, successful ones) on some scale of radicalism will reveal two distinct clusters. Were there,

Incremental Improvement and Radical Re-engineering

A common assumption:

Assess (say) 50 representative change projects on a scale of radicalism, and they will come out something like this.

But why should that be so?

This kind of distribution seems just as plausible.

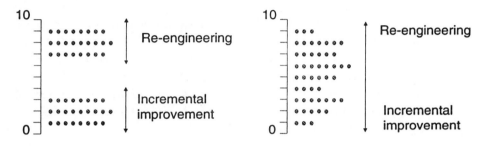

on the other hand, some other distribution of items across the scale, all three of the positions above would be undermined.

What is the truth? Rank filmstars by fame, countries by economic prosperity, or species of animal by size, and there won't be two neat clusters. Anyone who holds the view that changes in business processes ranked by radicalism do form two clusters ought to justify it. But neither Davenport nor anybody else adduces any evidence.

Moreover, there are good arguments against the sharp distinction. In his opening chapter Davenport explains that the 'clean slate' characteristic of re-engineering stands for finding the best means of accomplishing an objective, *regardless of how that has been done in the past.* But later in the book, he gives examples of constraints, rooted in current circumstances, that often make it impractical to design an ideal new process.

As he points out, a design based on a new case-manager concept may seem attractive, but if the staff who operate the present process are judged unsuitable for the new roles, that constraint may block the idea. Or you may envision new types of insurance policies handled by radically new processes, but the necessity to support a legacy of policies issued under existing processes may thwart you. And again, it may be pointless to design an ideal new process, regardless of how it has been done in the past, if existing IT arrangements form a decisive impediment to rapid change.

Reviewing such examples Davenport urges that constraints be recognised and allowed for rationally. This is sensible, but it abandons the stipulation that a true re-engineering project must proceed from a constraint-free clean slate. On a few projects there may be no constraints to speak of; on others one constraint may be overwhelmingly powerful; but on others, probably most, there may be several constraints of varying weight and influence — some, though perhaps not insurmountable in themselves, exacting a price in the calculus of tradeoffs.

In his later article with Stoddard, Davenport seems to recant, and says that in practice clean-slate change is rarely found, a view that considerably reduces the force of the opening chapter of his book.

What Does 'Radical' Mean?

Or take a less demanding criterion than the 'clean slate': the re-engineering projects are the ones with radical aims and results, while the less radical projects don't count as re-engineering. The following four examples from a publishing company might reasonably be located at different points on any scale of radicalism:

● a better design for taking and scheduling magazine adverts, with 30 activities on the workflow chart instead of 90, and extra advice to advertisers on marketing strategy; staff savings 40%; advert cycle-time reduced by 60%;

● a better design for taking and scheduling magazine adverts, with 50 activities on the workflow chart instead of 90, and without any extra advice feature; staff savings 40%; advert cycle-time reduced by 40%;

● a better design for taking and scheduling magazine adverts, in which only the (terribly inefficient) procedures for setting up a new advertiser's account are entirely changed, while the rest of the process is tidied up in some obvious ways; staff savings 20%; advert cycle-time reduced by 20%;

● exactly the same process for taking and scheduling magazine adverts as before, except that all the screens for keying in data are redesigned to be easier to use; staff savings 10%; cycle-time reduced by 10%.

The first example is presumably radical re-engineering. Is the second radical enough to count? Probably. What about the third? Probably not. But different people will have different opinions about what counts as a radical change, and there is really no way of settling any argument. Moreover, in one publishing company even the third might be the most radical development for decades, whereas in another it would be staid compared to other changes under way.

▼ Recognising degrees of radicalism helps generate options. Two, perhaps three, or even all four of the above could be credible options for a given publishing company. If that is so, then certain extreme statements about re-engineering can only be treated as exhibits for the office Wall of Shame: 'You can choose to re-engineer, or you can choose to go out of business' (Morris and Brandon); 'Forget what you know about how business should work — most of it is wrong!' (cover of the Hammer and Champy book); 'Be Great or Be Gone' (Texas Instruments marketing brochure).

Of course, silly assertions are easily found on many other topics too: total quality management, green accountancy, the learning organisation etc. Many of them are vulnerable to critical thinking, along the lines: 'Is there really an either/or choice here, or are there many degrees, and thus many options?' ▲

Using the Rhetoric

The above seems to suggest that arguments about whether re-engineering is or is not inherently better or more significant than other approaches, or whether any given instance is truly re-engineering or not, are futile. In any particular situation you should simply apply whatever degree of radicalism is appropriate, given all the circumstances.

That is a rational point of view, but perhaps factors other than reason can come into play. For anybody involved in managing a particular re-engineering project, the crucial question is not whether Hammer and Champy's bombast or Davenport's sharp distinction contain any wisdom, but what use of rhetoric *in this particular project* will be advantageous.

In *Beyond the Hype, rediscovering the essence of management* (Harvard Business School Press, 1992), Robert G Eccles and Nitin Nohria remark that the familiar rhetoric about revolutionary change, post-industrial society, knowledge workers and so on can easily be found in management writings *throughout the last 50 years*. Rather than finding this cause for cynicism, the authors argue that 'rhetoric itself is a kind of action.' They advocate that, as a manager, you should therefore improve your rhetorical skill.

Eccles and Nohria seem too timid to draw the logical conclusion that you should coldbloodedly devise the rhetoric to suit your own objectives. Take a project whose aim is to improve the process of handling insurance claims on household policies, for burnt carpets and the like:

● You may want to raise the stakes by talking about valour and obliteration and smashing china and such things — not just once, but as deliberate policy in every meeting and document. Maybe that will be helpful in motivating people and making them more imaginative and liberated, so that designers conceive ingenious ideas for handling claims, and workers welcome the changes.

● On the other hand, the designers may become so psyched up that they veer outside the defined scope of their project (eg recommend taking over a carpet manufacturer), or put forward outlandish proposals (eg for outstandingly generous claims payments) that don't stand up to scrutiny. Or the users may be bitterly disappointed that the new process, when it comes, is not as spectacularly different from the old as the emotive language implied. Or again, perhaps if you pitch things too high, people will respond with coarse obscenities, like veterans of the trenches addressed by a conceited general.

• But, then again, if you merely point out that an efficient claims process is a good thing for an insurance company to have, and invite people to do their best, perhaps they won't summon up the verve and zeal essential to overcoming the organisation's innate conservatism.

This suggests that rhetoric needs to be consciously, even cynically, chosen. In exhortation about clean slates, a revolutionary manifesto, and being great or being gone, truth and logic and relevance to verifiable facts are hardly considerations of the first importance. The choice of rhetoric must be based on psychological considerations, and, as the options above suggest, you need to strike the right balance for the situation.

CONNECTIONS

The place of IT people in re-engineering teams, one of the example topics above, is part of the more general topic of the impact of IT factors on re-engineering. This is surveyed at length in Briefing 8.

If this book had accepted the assumption of a sharp distinction between change by re-engineering and all other change, that would have had repercussions for the very definition of re-engineering, one of the topics of Briefing 2.

This briefing is concerned above all with dissecting the characteristics of change-management factors, by contrast with directly-rational factors. The argument presented is that change-management factors can, in ways hard to generalise about, affect any of the decisions discussed anywhere in Briefings 3-10.

Appendix: Suppliers of Software Products

This provides supplier details for software products mentioned in Briefings 6 and 7. As far as possible, a European, English-language contact is given.

ABC Flowcharter	Micrografx Inc, 1303 Arapaho, Richardson, TX 75081, USA.
Action Workflow Analyst	Action Technologies Inc, 1301 Marina Village Parkway, Alameda, CA 94501, USA.
ANSWER:Architect **ANSWER:Cabe**	Sterling Software, 1 Longwalk Rd, Stockley Park, Uxbridge, Middlesex, UB11 1DB, UK.
Bachman/Analyst	Bachman Information Systems, 8 New England Executive Park, Burlington, MA 01803, USA.
BPwin	Logic Works Inc, 1060 Route 206, Princeton, New Jersey 08540, USA.
Business Design Facility (BDF)	TI Information Engineering Ltd, Wellington House, 61-73 Staines Road West, Sunbury-on-Thames, TW16 7AH, UK.
CorelFlow	Corel Corp, 1600 Carling Ave, Ottawa, Ontario, K1Z 8R7, Canada.
EasyCASE	Evergreen CASE Tools, 8522 154th Ave, Redmond, WA 98052, USA.

Intellidraw	Aldus Corp, 411 First Ave South, Seattle, WA 98104, USA.
ithink	Cognitus Systems Ltd, 1 Park View, Harrogate, North Yorkshire, HG1 5LY, UK.
MacFlow	Mainstay, 591-A Constitution Ave, Camarillo, CA 93012, USA.
MAXIM	Sterling Software, 1 Longwalk Rd, Stockley Park, Uxbridge, Middlesex, UB11 1DB, UK.
Object Management Workbench	IntelliCorp Ltd, Unit 6, Bracknell Beeches, Bracknell, Berkshire, RG12 7BW, UK.
ProcessWise WorkBench	ICL, Waterside Park, Cain Rd, Bracknell, Berks, RG12 1FA, UK.
RADitor	Co-ordination Systems Ltd, 3c Cornbrash Park, Bumpers Way, Chippenham, Wiltshire, SN14 6RA, UK.
SES/Workbench	Scientific and Engineering Software UK Ltd, Corinthian Court, Milton Park, Abingdon, OX14 4RY, UK.
System Architect	Popkin Software, 11 Park Pl., Room 1516, New York, NY 10007, USA.
Taylor II	F&H Simulations bv, Spoorlaan 424, 5038 CG Tilburg, the Netherlands.
Visio	Shapeware Corp, 1601 Fifth Avenue, #800, Seattle, WA 98101, USA.
WITNESS	AT&T ISTEL Ltd, Highfield House, Headless Cross Drive, Redditch, Worcs, B97 5EQ, UK.

Index

This index has five sections: 1. Authors; 2. Concepts and Topics; 3. Examples; 4. Information Technology; 5. Reasoning.

2. Concepts and Topics

3. Examples

4. Information Technology

5. Reasoning